Published by:
Gibson Applied Engineering and Technology, Inc (GATE)
Houston, TX
www.GATEINC.com

ISBN-13: 978-1517476755
ISBN-10:1517476755

EDITOR

Howard Duhon, P.E.

CONTRIBUTORS

Lindsay Anderson

Jorge Garduño

Janet Elias, P.E., Ph.D.

TABLE OF CONTENTS

Preface

References

Appendix – Example Procedure

PREFACE – SOPS AS A CATALYST FOR CULTURE CHANGE

What Is!

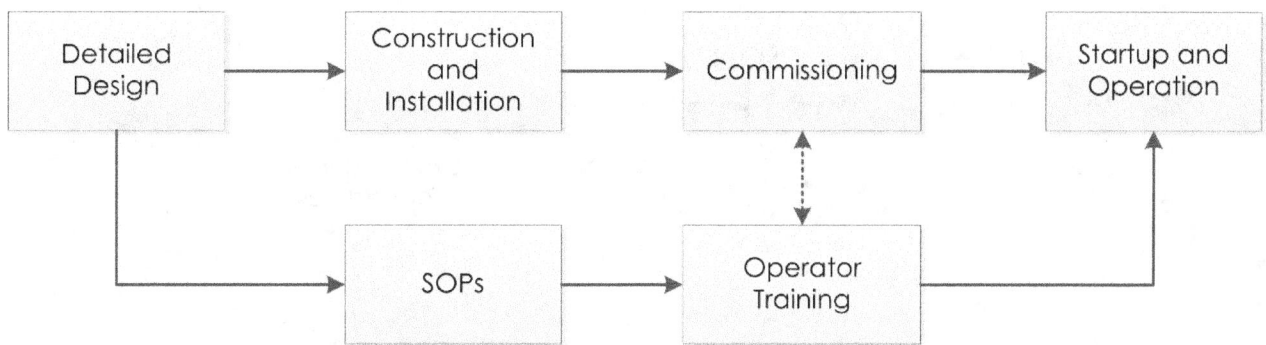

SOPs are important for the safe and effective operation of industrial plants, including oil and gas facilities. But SOPs are frequently developed late in the project, after the design is completed and construction is well underway. And they may be used for little other than operator training. Following startup, they often end up on a shelf collecting dust. Used this way, SOPs have little influence on either the design or the operation of the facility.

What Could Be!

SOPs can be so much more. They can be both an integral part of the design process and a catalyst for culture change in the industry. As part of the design process, there are several advantages to the SOPs:

- If generated early enough, they can be a powerful tool for finding and correcting operability issues in the design.
- They allow us to design the facility to be operated, rather than having to figure out how to operate a facility after it has been built.
- If generated early and in coordination with the control systems design effort, they provide guidance for the HMI design. They allow us to design the HMI for the way the facility will be operated.
- Effective SOPs can dramatically simplify the development of final commissioning and initial startup procedures.

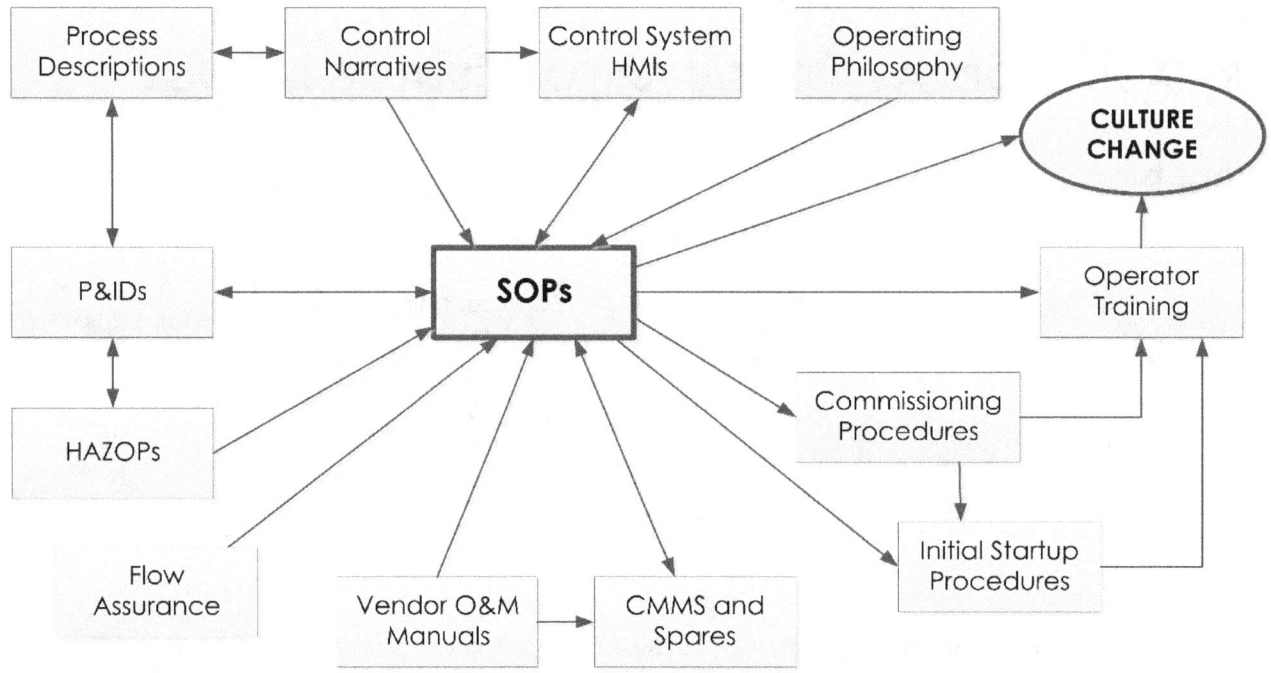

Changing Safety Culture

In January 2009, US Airways Flight 1549 piloted by Captain Chesley "Sully" Sullenberger and First Officer Jeffery Skiles hit a flock of geese. The plane lost power. Skiles reached for the SOPs. From the moment of the impact until the plane landed in the Hudson River, he diligently and rapidly applied procedures in his attempts to restart the engines.

The airline industry has dramatically decreased the frequency and impact of accidents by creating a culture of following procedures, typically in the form of checklists. They do this for normal events – takeoff and landings – and in response to emergencies. Airline pilots and other personnel have learned to rely on SOPs in all types of operating situations.

It is hard to imagine every having such devotion to SOPs in the oil and gas industry, but it is something we should strive for. Effective SOPs that are diligently used by operators may be the driver for the next step change in process safety in the industry. But this cannot reasonably be accomplished unless we start writing procedures that are more user-friendly - procedures that the operators will actually *want* to use.

Organization of the Book

Section 1: Introduction and Important Concepts

> This section summarizes the important concepts that guided the development of the GATE SOP process. Includes learnings from projects and from literature.

Section 2: The Procedure Writing Process

> In this section, we describe the GATE SOP process in detail. The GATE process is different. We use a phased approach that enables a much earlier start to procedure writing. SOPs written early enough can serve many functions as described above.

Section 3: Procedure Reviews and Operator Training

> This section describes methods of reviewing procedures and of training operators.

Appendix

> An example procedure is provided.

SECTION 1: INTRODUCTION AND IMPORTANT CONCEPTS

CHAPTER 1 – INTRODUCTION TO THE GATE SOP PROCESS

This book describes the GATE process for generating Standard Operating Procedures (SOPs) and other similar procedure documents. The process is based on over a decade of learning from writing SOPs for projects as well as extensive literature study.

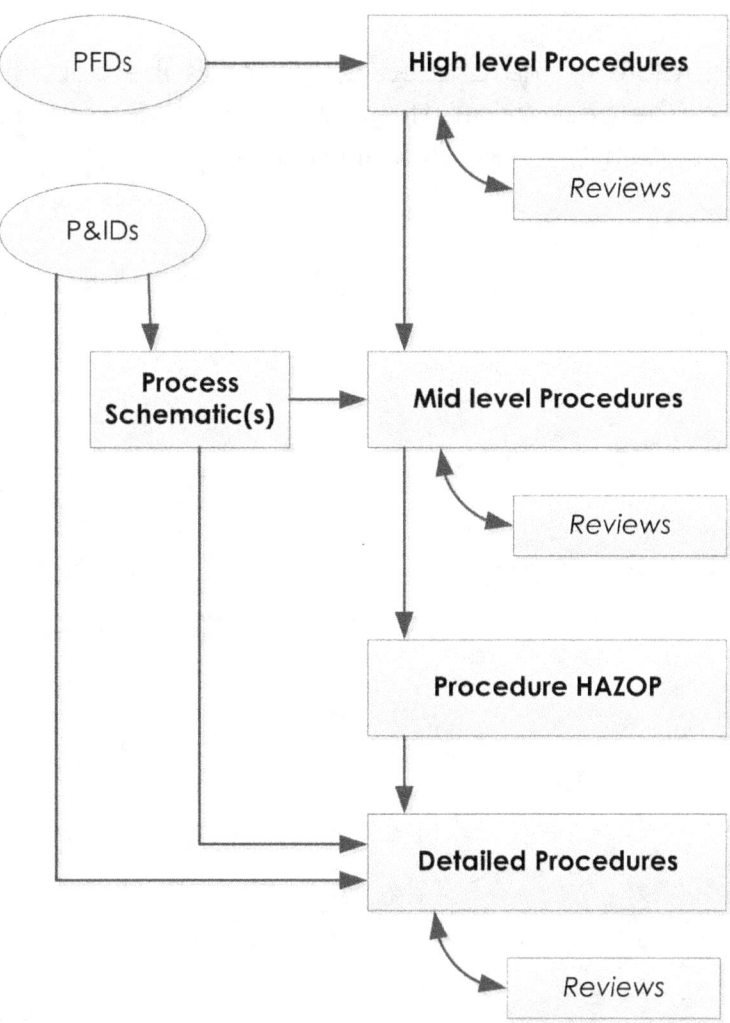

The key feature of the process is phased development. It is relatively common practice in the industry to generate detailed procedures straight-away. There are several reasons for using a phased approach. Detailed procedures are difficult and time consuming to write and they are difficult to review. Because they are difficult to review, hidden errors embedded in detailed procedures may not be found. And when the errors are found, it will often be difficult and time consuming to correct them.

Conversely, High-level and Mid-level procedures are easy to review and easy to change. The important decisions are made at the Mid-level. Detailed procedures are then easier to write and contain fewer errors.

The most important reason for using a phased approach is that it allows us to begin writing the procedures very early in the design process. High-level procedures can be developed from the PFDs.

Another important feature of the phased approach is the opportunity to perform a procedure HAZOP at the appropriate stage: Mid-level. Mid-level Procedures contain enough information to identify the significant hazards.

CHAPTER 2 - LESSONS LEARNED FROM PROJECTS

We have been writing procedures including SOPs, Commissioning Procedures, and Initial Startup Procedures for over a decade now. Many projects have taught us valuable lessons. Our current process is largely the result of the following important learnings.

SOPs for Nigeria FPSO – 2002
We had significant involvement in systems engineering tasks for much of the project. We started the SOP development late in the project. The SOP development was a humbling experience as we found numerous design flaws. But since the facility was already nearly complete and running behind schedule we could not make any changes to the design.

Nigeria Offshore Pipeline to LNG Facility - 2004
We attended the HAZOP for a project that added an incoming gas pipeline to an LNG facility. After the HAZOP, we wrote High-level startup and shutdown procedures and identified a serious operability problem. A HIPPS system was installed at the pipeline tie-in. After a trip of the pipeline HIPPS system, the pipeline would be allowed to pack (increase in pressure) prior to a trip of the production facility. It then would have taken several hours to equalize pressure across the HIPPS prior to restarting the line because only a small equalizing valve had been provided. We caught it in time - addition of a startup choke solved the problem.

Learning 1: Write the operating procedures early enough to impact the design. Seek to design kit to be operated rather than trying to figure out how to operate kit after it has been built.

SOPs for Nigeria FPSO – 2002
After completion of the detailed procedures, the operating company conducted a Procedure HAZOP. The facilitator had us first combine related detailed steps into a node and then asked the HAZOP questions about the combined steps. For example, 6 steps might result in a node called "Start the Booster Pump on Recycle".

Combining the related detailed steps amounted to creating Mid-level procedure steps. It seemed clear as we were doing this that the HAZOP should have been done earlier on Mid-level procedures, before all the detailed steps had been written. But there was no point on that project in which we had a set of Mid-level procedures; and hence no such opportunity. We had written the detail procedures straight-away.

Learning 2: Write the procedures in stages. Perform the HAZOP at a Mid-level stage prior to writing the detailed procedures. Following that learning our first thoughts on a procedure process were:

1. High-level Procedures
2. Mid-level Procedures
3. Procedure HAZOP
4. Detailed Procedure

GoM Subsea Tieback Procedures – 2005

Our first attempt to implement the new High-Mid-Detailed process was disappointing. Though we clearly articulated the phased approach to the team performing this project, we could not get them to follow it. The lure of detailed procedures is strong in engineers. In effect, they tended to write the detailed procedures first and then extract the High-level and Mid-level steps. (Not what we had in mind!)

Learning 3: Engineers have a strong tendency to gravitate to the details early. Significant training, discipline, and management control are required to establish the phased approach.

South America Gas Development – 2009

This project involved addition of a gas development platform integrated with an existing oil production platform. Three sets of P&IDs were involved. The startup was complex with many options. One of our first actions was generation of an overall schematic that showed, on one page, everything we needed to know about the existing and new platforms for assessing the startup options. The resulting drawing is shown below. It summarized content from 42 P&IDs from three drawing sets.

This was very successful. Even printed on and 11x17 sheet it was possible to read equipment tags. The schematic was used for all planning and procedure writing except for some of the detailed step development very late in the process.

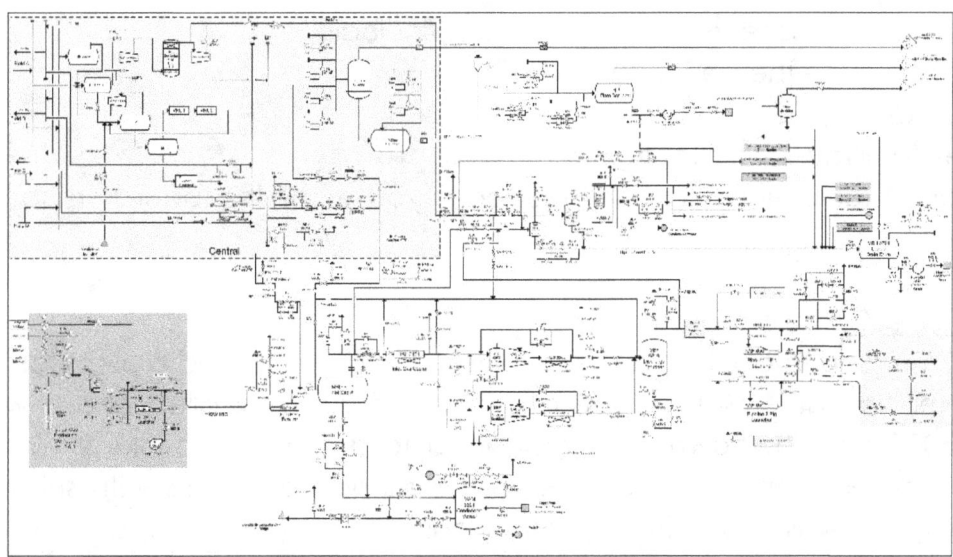

Learning 4: One of the reasons that engineers gravitate early to the details is that the governing documents for SOP development are the P&IDs which are very detailed. To avoid this tendency, GATE generates overall system schematics prior to writing the procedures. Then we put the P&IDs away until we need them for the detailed procedures. These schematics, that show the entire project on one or two sheets, are detailed enough for all the High-level and Mid-level procedure development and for most of the detailed procedures.

GoM Subsea Tieback – 2007

For this subsea tie-back startup, we wrote detailed procedures (over 300 steps). We trained the operators. We attended a toolbox talk prior to the startup. The Control Room Operator (CRO) had the procedure in his hand during the toolbox talk, but he never looked at the procedure during the startup.

And yet the procedures were valuable. The operator in question had been part of the effort to generate the procedures and knew them well.

During procedure development we found that the flow assurance plan had to be extended to the lower rates and lower temperatures experienced for the initial startup. We also found that subsea chokes were too large to control flow at the low startup flowrates requested by the Reservoir Engineers. Adequate solutions were found for both issues.

Learning 5: Procedures serve multiple functions including:
1. They validate the design
2. They identify risks
3. They facilitate operator training

Management intervention may be needed if you want the operators to implement them line-by-line.

GoM Tieback – 2007 and many others

In the case mentioned above, the reservoir engineers requested a 50 psi drawdown at the initial well flowrate. But the first choke step for the installed choke would have yielded a 500 psig drawdown. We managed to get a new choke insert in time for startup. On several other startups, we've experience problems with subsea chokes, topsides arrival chokes, water injection flow control valves, and separator level control valves.

Chokes and control valves are frequently oversized for normal operation and are so large as to be nearly useless at startup conditions.

Learning 6: Just because you have a control valve doesn't mean you have control. Control valves and chokes are frequently oversized. Control valves and chokes, at least the important ones, need to be checked against the full range of operating conditions.

TLP with Subsea Production – 2009

On startup of a subsea field with a looped flowline, we were choking heavily on topsides because the subsea chokes were too large to effectively control the wells at low flowrates. Because we were choking at topsides, switching a well from one flowline to another was a tricky operation. This is because switching a well from one flowline to another significantly changes the flowrate in both flowlines and so dramatically changes the pressure drop through both topsides chokes (pressure drop is a function of flowrate squared).

This was a delicate operation and was carefully detailed in the procedure. But the first time a well was switched, the operator skipped the important equalization steps. This caused dramatic pressure swings in both flowlines and tripped the entire platform.

It is tempting to blame this on operator error, but that would be a mistake. While the procedures were accurate, the steps required were not intuitively obvious. We should have highlighted, in the procedure, the fact that switching under startup conditions was different that switching would be later in field life`.

Learning 7: It is not good enough to write accurate procedures. We must write procedures that are also likely to be implemented accurately. Potential for human error can be minimized by applying learnings from human error research (see Chapter 3 and 11).

West Africa Platform – 2013

Procedures are system based (separate procedures for the Separation System and the Oil Export System for instance). But startups and shutdowns cross system boundaries. The interaction between systems cannot be effectively captured within the system-based procedures. On this project we found that Mid-level procedures could be arranged to provide a comprehensive startup procedure.

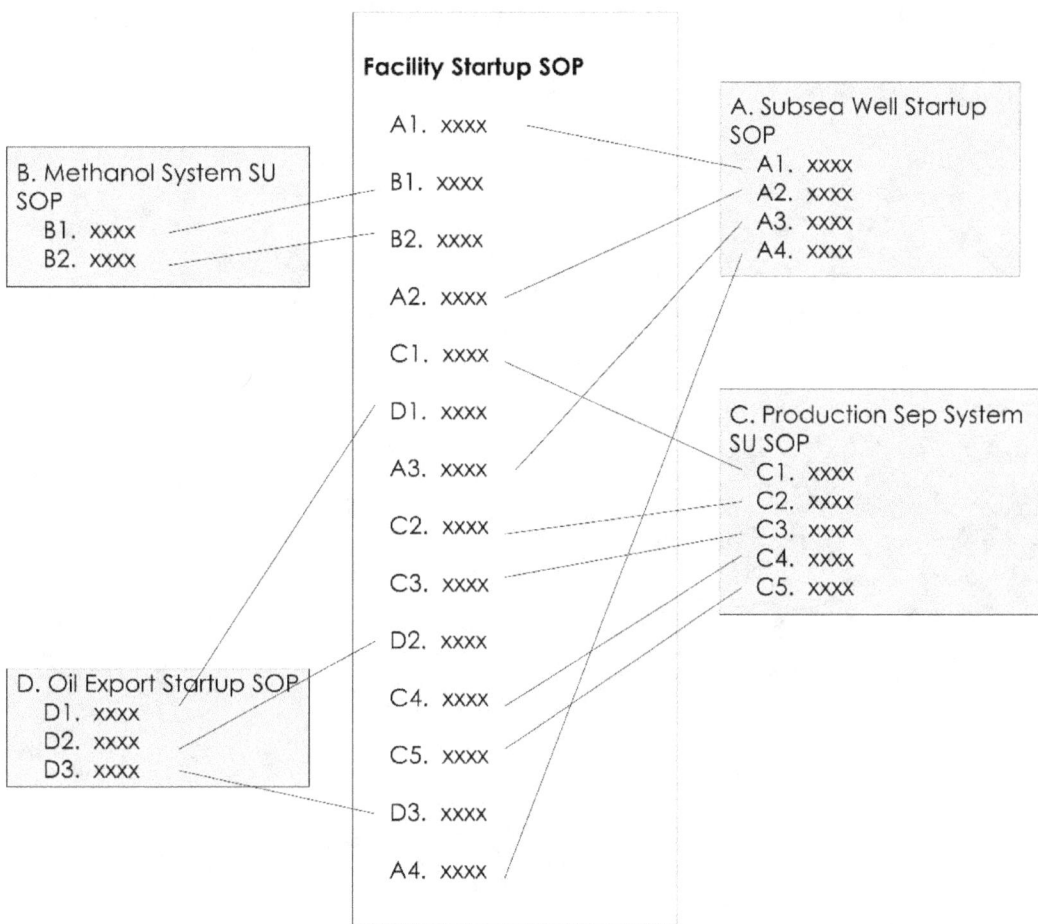

Learning 8: Mid-level steps from each of the system-based procedures can be combined to yield overall startup and shutdown procedures.

CHAPTER 3 – LESSONS LEARNED FROM LITERATURE

Summary Literature Sources

GATE has been obsessed for some time with learning how to better construct procedures. That obsession, and some serendipity, has taken us to several fields of study for insight. This section contains important learnings including:

1. Naturalistic Decision Making (NDM)
   We have studied decision theory in general looking for insights. NDM, the study of how people under stress make decisions, has been particularly useful. (see Klein, 1998)

2. Checklists / CRM
   There is much more to Crew Resource Management (CRM) than checklists, but checklist development is one of the important insights. (see Gwande, 2009)

3. Human Error Research
   It is not good enough to write accurate procedures. Accurate procedures can be implemented incorrectly. Human error research provides guidance on how to minimize the potential for error in following procedures. (see Reason, 1990)

4. HROs
   Highly Reliable Organizations (HROs) are organizations that operate in hazardous industries, but manage to have fewer incidents than other companies operating in those industries. (See Weick and Suttcliff, 2007)

5. Structured Writing (such as Information Mapping)
   Structured writing instructions (chunking, keeping is short, simple instruction, integrated graphics) seem tailor-made for procedure construction. SOPs are highly structured documents.

6. Use Cases
   Computer programmers have a very difficult task. Whereas we can expect an operator to adapt in the field to unexpected events, programmers must anticipate everything that can go wrong and program the computer to respond appropriately; whether that means recovering or failing gracefully. This structured approach is a powerful way to inspect for things that can go wrong. (see Cockburn, 2001)

7. API RP 75

    API RP 75 has only ¾ of a page devoted to operating procedures. At first glance
    it looks inadequate, but if you read that page carefully and thoughtfully, there is
    a lot of good advice.

NATURALISTIC DECISION MAKING - Decision Making Under Stress

The classic decision model is:
     Gather data,
     Identify objectives,
     List alternatives,
     Select the best alternative

But this does not accurately describe the way people make decisions under stress or
time pressure.

The graphic below is a simplified schematic of Recognition Primed Decision Model
(RPD) which describes how people make decisions under time pressure. (Klein, 1998)

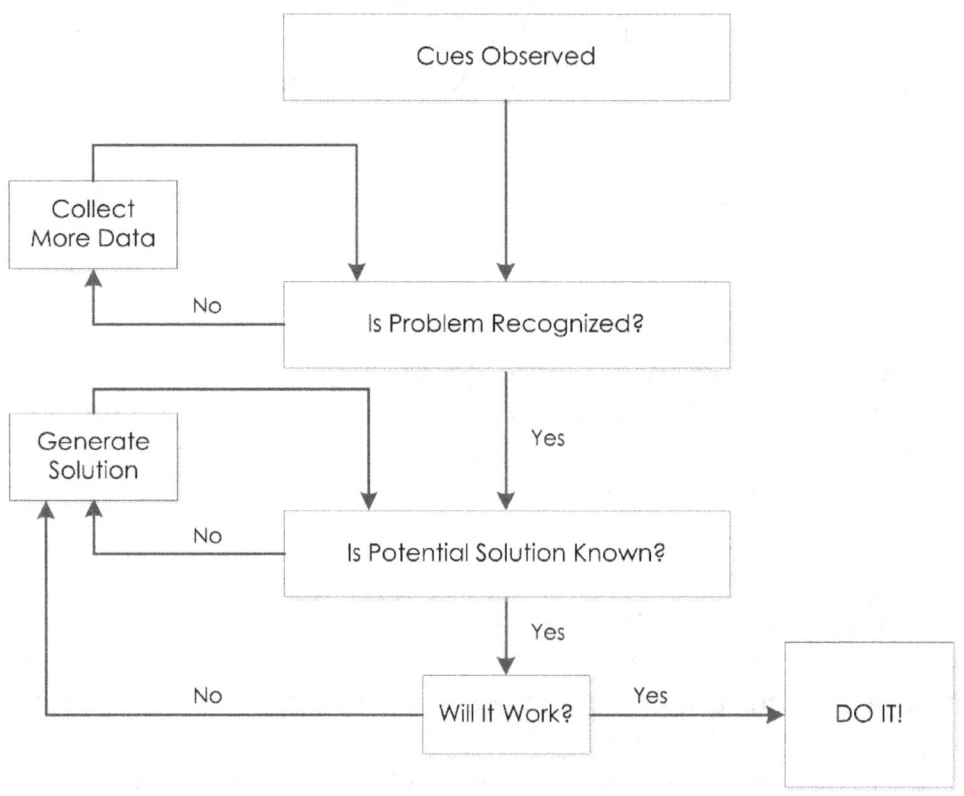

RPD Description

1. The process begins with observed cues that there is a problem.
2. The expert witnessing the cues asks himself "Do I recognize this problem?"
3. If he does not recognize the problem, he will gather more data in his attempt to diagnose the problem.
4. If he does recognize the problem, he next asks himself "Do I know a possible solution?".
5. If the answer is "No", he must do some problem solving to attempt to identify a solution.
6. Once he has identified a potential solution he asks "Will it work?"
7. If the answer is "No", he must cycle back to problem solving to identify another possible solution.
8. If he believes the solution will work, he implements it.

Sources of Stress

It is easy to see, in the figure above, what the sources of stress are. If an operator, upon seeing cues:

- Immediately recognizes the problem
- Knows a potential solution
- Is confident that the identified solution will work
- Implements the solution successfully

He/she will feel little or no stress. Stress comes from:

- Not understanding what is going on
- Not knowing how to fix it
- Running out of time

Requisite Skills – Pattern Recognition and Mental Simulation

This RPD model is employed by experts. Experts possess two skills that allow them to solve novel problems:

1. **Pattern Recognition (Situation Assessment)**
   Seeing the problem early is key. The ability to recognize that something is wrong from the early cues and the ability to quickly identify possible causes and possible solutions is critical to effective behavior under time pressure.

2. Mental Simulation Ability

Once an expert is confident that he/she understands the problem and has identified a possible solution, the next task is to predict the effects of implementing the solution. To evaluate a solution, the operator must predict the behavior of the system with and without the proposed solution.

Novices do not possess the tools of pattern recognition or the mental library of possible solutions. Furthermore, they do not have the systems knowledge necessary to mentally simulate potential solutions "on-the-fly"; hence, novices are forced to be more structured, to collect more data, and to list and evaluate alternatives before making a decision.

CHECKLISTS

In October 1935, the flight competition for the next US long-range bomber was Boeing's to lose. The plane Model 299, later called the B-17, crashed on takeoff. After the crash, it was labeled "too difficult to fly" and Boeing lost the contract. The ultimate solution for the B-17 was not redesign of the aircraft. Instead, checklists were developed. Boeing went on to sell 13,000 B-17's. (Gwande, 2009)

Experts face two main difficulties in a complex environment:
1. Fallibility of human memory and attention. In particular, mundane, routine matters are easily overlooked in the strain of more pressing events.
2. People lull themselves into skipping steps even when they remember them. Some steps don't always matter.

Checklists can provide protection against such errors.

Three types of problems exist:
1. Simple. Like baking a cake from a mix. Just follow the instructions and everything will come out fine every time, even for a beginner.
2. Complicated. Like firing a rocket. There is no straightforward recipe. Expertise is required and frequently a multi-skilled team, but once you get it right you can do it repeatedly.
3. Complex. Uncertainty is high. There is no 'correct' way. Even if you've done it well before there is significant risk of failure the next time.

Checklist must be attuned to the type of problem.

For complex problems, decision making must be with the experts on the ground not commanded from above.

A good checklist is:
- Precise
- Efficient
- To the point
- Does not try to spell out everything. It is not a complete procedure. It is a reminder of the <u>most critical steps</u> that even an <u>expert might miss</u>.

Checklists must be short – no more than 10 items (5 to 9 usually suggested). A checklist should fit on one page. The wording must be simple and exact.

Pilots use checklists for two reasons:
1. They are trained to do so.
2. They work.

A checklist can be either
- DO-CONFIRM. At a pause point the team checks that required actions have been done satisfactorily before continuing.
- READ-DO. Team members read each step as it is performed (following a procedure).

Example: Landing UA Airways Flight 1549 in the Hudson:

Co-pilot Jeffrey Skiles and Captain Chesley B. "Sully" Sullenberger had never flown together. Prior to the flight they:
- Introduced themselves
- Met the cabin crew
- Ran through checklists, and
- Talked about the flight plan.

In this process they transformed themselves into a team.

Skiles was flying the plane on takeoff. They hit a large flock of Canada geese and lost both engines.

Sullenberger immediately decided to fly the plane and simply announced "My aircraft." Skiles replied "Your aircraft." (standard communication minimizes probability of miscommunication.)

They did not discuss what to do next; there was no need. Sullenberger looked for a place to land. Skiles tried to restart the engines following SOPs/checklists developed for the purpose. He managed to try a restart on both engines, something investigators considered remarkable.

Meanwhile the cabin crew got everyone prepared for the water landing. Working together they got everyone out in three minutes according to plan.

After the landing, Skiles remained in the cockpit to go over the emergency landing checklist while Sullenberger checked the plane. He walked the aisle to make sure no one was left behind and then exited.

Because of incredible training and a culture of adhering to rules, they effectively performed something most crews never experience.

HUMAN ERROR

Accurate plans may be implemented incorrectly due to human error. In *Human Error* James Reason identifies many causes of error (Reason, 1998). Below is a summary of his observations relevant to the potential for operator error in following a plan or procedure.

Pre-conditions are likely to be incomplete or not done if or when:	
1	Following maintenance/repair the status of equipment may not be as expected. *For instance: slip blinds left in place, equipment not powered, safety systems bypassed, instruments on manual, etc.*
2	Checklist does not follow logical progression. *For instance: valves in a line required to be in a given position should be listed in line order to facilitate checking.*
3	Required conditions prior to proceeding to next step are ambiguous
Errors in implementing steps in a procedure are most likely:	
4	If a step is not obviously cued by the previous step it is likely to be skipped
5	If there are too many steps, steps in the middle are likely to be skipped
6	If a single step is complicated and has sub-steps, slips and lapses are likely
7	A step which extends over long period of time without intervening action is prone to distraction. The operator may not notice cues to continue to the next step. And the operator may forget where he is in the process and may skip steps or repeat previously accomplished steps.
8	Steps which occur after the main goal is achieved are likely to be skipped.
9	Interruptions from SIMOP's (simultaneous operations) can cause you to lose your place.
Latent Errors – Problems waiting to happen	
10	Look for latent errors caused by skipped steps. If a step can be skipped without impacting further steps or causing immediate upset to the process, then it is likely that failure to perform the step will not be caught. Somehow reinforce the need to complete the step.

Another perspective on reducing human error potential is available from human factors engineering. OGP Report No. 454 (International Association of Oil & Gas Producers, 2011) is a good summary. Insights for assessing SOPs include:

Importance
> How important is this task?
>> Personnel safety risk?
>> Process safety risk?
>> Environmental risk?
>> Asset/production risk?

Complexity and Novelty vs. Operator Knowledge, Experience, and Skill
> Complexity - How complex/difficult is this task? Are there known problems with the process?
> Novelty - How frequently is it performed?
> Knowledge & Experience - How skilled are the operators in performing it?

Time and Attention
> How time-consuming is this task?
> What fraction of the operator's time and attention is devoted to this task?

Coordination and Interface Issues
> How many people are involved? Are roles and responsibilities clear?
> What communication is required if the task extends across shift changes?

HIGHLY RELIABLE ORGANIZATIONS (HROs)

Highly Reliable Organizations (HROs) manage to operate in dangerous industries with fewer than their share of accidents (Weick and Sutcliff, 2007). Weick and Sutcliff claim that this is because they are better at anticipating problems and containing damage. Consideration of the properties of HROs can help us generate better procedures. HROs do five things well:

1) They are preoccupied with failure rather than success

It is easy to become complacent because serious accidents are rare. HROs guard against complacency by focusing on potential failure and always being on the lookout for cues that something is going wrong.

- Failure detection is most likely if everyone is on the lookout and expected to identify and communicate abnormalities.
- Failure detection is improved if there is a detailed list of expectations. Failures are detected if the actual results are different than what was anticipated.

- Failure detection is more likely if the team is integrated and players have an overall systems view. Where people live in 'silos' failure detection is impaired.
- Reporting failure is key. People must feel comfortable reporting their errors. The best HROs increase their knowledge base by encouraging and rewarding error reporting.

Cues that something is going wrong can be hard to distinguish from "noise". HROs are mindful of this and do not readily dismiss cues of potential problems. Everyone is encouraged to identify and report potential problems.

2) They are reluctant to simplify

It's typical to simplify categories such as "make or buy", "friend or enemy", "profit or loss". But the world is complex and simple categories cause us to overlook important information. The classic example of this is the "empty" drum of gasoline. "Empty" suggests "safe", but an empty drum filled with gasoline vapor may be more dangerous than a liquid-filled drum.

A common simplification is to trust drawings. Drawings may be outdated and may have omitted details that may be important in specific contexts. This is certainly a consideration for the simplified sketches included within procedures. Another common simplification is to assume that a kit will work even if the operating conditions are far from the design conditions. Control valves, meters, chokes, rotating equipment, columns, etc. all have turndown limits and other limits.

3) They are sensitive to operations

Managers at HROs keep abreast of what is happening "on the ground."

Sensitivity to operations means paying attention to what's actually happening vs. what's supposed to be happening.

The primary threat to sensitivity to operations is the engineering culture that puts a higher value on knowledge that is hard and quantifiable (engineering) rather than experiential knowledge (operations).

Another threat to sensitivity is the tendency for routine actions to be implemented mindlessly. "Routine" and "mindless" are not synonyms. Routine actions can be done mindfully. JSAs and work permits can serve to create mindfulness in routine actions, but often routine tasks are exempted from work permits.

A third threat to sensitivity to operations is the tendency to overestimate the soundness of systems and processes. This can cause people to learn the wrong lesson from close calls. A near miss can be interpreted as a sign that something is wrong or as evidence that the system is resilient.

Another threat is confirmation bias: When we succeed we attribute the success to skill even if it was largely luck. Success narrows perspectives, breeds overconfidence, and reduces acceptance of other points of view.

4) They commit resources to being resilient

HROs plan for the unexpected. Minor problems don't get out of control because there are plans in place to deal with them. Also, people on the front line know that they have the authority to do whatever is necessary to deal with a problem.

The world is more complex than our models. Our plans for the future are inevitably simplifications of how the future will unfold. Things may go wrong in many unexpected ways. Our plans create expectations, and expectations bias our interpretation of results. We are biased to see what we expect to see. When people impose their expectations on ambiguous cues, they fill in the blanks. Slight deviations from expectations are smoothed over.

Errors are inevitable. We need to be as concerned with cues as we are with prevention. Mitigation prevents an outcome from getting worse.

Resilience occurs when the system continues to function despite failure of some of its parts (degrading gracefully). Organizations often respond to unforeseen events by imposing rules to prevent the disturbance in the future. These rules often reduce flexibility and make the organization less resilient.

5) They give deference to expertise rather than bureaucracy

Emergencies must be managed by the people with the expertise to understand and deal with the problem. During an emergency, management must yield to the experts.

The first persons to detect an error are frequently lower level employees. Management gets the news late - and it is probably sugarcoated.

In an emergency situation only people on the front line and/or specifically trained to handle the emergency can respond effectively. Only they have the process knowledge and access to data required to make informed choices during the emergency. They must be free to operate without management interference and without having to deal with bureaucratic issues.

STRUCTURED WRITING

Methods such as Information Mapping™ address the structure of a document. The general idea is to structure a document that contains a large amount of information in a way that makes that information easily available to the reader.

A document should be written specifically to answer reader's questions:
1. What is the main point of the document?
2. Why me?
3. Why now?
4. What should I do?
5. What do I need to know in order to do it?

Main Principles

Some of the important principles of structured writing are:

1. Chunking
   Information should be grouped into small manageable units (chunks). The document should have a limited number of chunks. The general advice is to follow the 7±2 rule and have less than 9 chunks. Chunks may be organized into hierarchies.

2. Labeling
   Each chunk should be labeled for easy reference and scanning. It should be easy for the reader to find the information he/she is looking for.

3. Relevance
   All the information contained in the chunk should be relevant to the one main point. For complex documents, organize relevant chunks into a hierarchy.

4. Consistency
   For similar subjects, each chunk should look a lot like other chunks; similar words, style, organization, etc.

5. Accessible Details, Integrated Graphics
    Put what the reader needs where he/she needs it. Required detail should be easy to find. Graphics should be used extensively and must be integrated with the text, not buried at the end of the document. If the same graphic is reference multiple times in the document, put it in each place rather than just once.

Types of Information

Information Mapping™ recognizes 7 information types as described below. Different types of information will generally be presented in different ways, but all incidents of a type should be similar (i.e. all facts/data presented in a similar format such as a table).

The information types are:

1. Procedure
    A list of steps leading to a result (operational 'how-to' level)

2. Process
    A means of accomplishing something, such as the means of obtaining management approval for a project budget. Distinguished from a procedure in that the steps may be tacit rather than explicit.

3. Structure
    Describes a physical, material object (printer, form, machine)

4. Concept
    Describes an idea, a concept

5. Principle
    A policy or rule specifying what is allowed, what is not allowed, what is required, etc.

6. Fact
    Data not subject to dispute.

7. Classification
    Sorting of chunks/units into classes.

Structured Writing Example

Consider the following paragraph copied from a book of stress among offshore workers:

> Although most individuals were in the 31-35 age group (26%), the distribution across all age bands was fairly consistent. Only 10% were less than 25 and 10 individuals (3.2%) were between 51 and 60. Contractors tended to be a slightly younger group; 17% were less than 25 years old, 23% were between 26 and 30, compared to operators where only 10% were less than the age 30. At the opposite end of the age range, 30% of operators were over 40 years of age, compared to 21% of the contractors.

Data, such as the age range percentages of operators and contractors, is difficult to communicate in paragraph form. The same data presented in a table, as shown below, is much easier to access.

Age Range	%		
	All	Operators	Contractors
< 25	10		17
26 – 30			23
< 30		10	40
31 – 35	26		
> 40		30	21
51 - 60	3.2		

USE CASES

Use cases are an invention of the computer software business (Cockburn, 2001).

A use case is a description of how an item, system or process will be used to achieve particular goals. Each goal typically requires a different use case. Use cases are typically nested with one use case referencing other use cases.

I find these very useful for specifying equipment items and systems. A description of how the equipment will be used (especially in abnormal situations) can avoid some embarrassing misunderstandings.

Use Case Contents:

Goal: Each use case has a specific goal which the primary actor wants to accomplish.

Actors: People, items, control systems, etc. that have goals and play roles.

Primary Actor: Person with the specific goal that the use case is built to describe.

Stakeholders: Everyone with a goal or interest in the outcome of the use case.

Triggering Event: Event that starts the use defined by the use case.

Success Guarantee: Definition of what needs to be accomplished for the lead actor to consider the use a success.

Minimum Guarantee: The minimum that the system must accomplish if success is not achieved. For engineering designs this is usually a safe result such as a system shutdown to safe conditions.

Main Success Scenario: A scenario which achieves success as the system is designed to operate (flawless operation).

Extensions: A list of things that could go wrong with main scenario steps and what should happen when things go wrong (corrections, mitigations).

The two most important components of a use case are frequently the main success scenario and the extensions. The table below is the main success scenario for the goal of "Withdrawing cash from an ATM".

Example: Withdrawing Cash from Automatic Teller Machine (ATM)

> **Main Success Scenario:**
> 1. Customer inserts card.
> 2. Machine verifies card and asks for PIN number.
> 3. Customer types PIN number and hits Enter Key.
> 4. Machine verifies PIN and asks customer to select transaction type.
> 5. Customer selects "Cash".
> 6. Machine asks customer to specify the account to withdraw cash from.
> 7. Customer selects account.
> 8. Machine asks customer to specify the quanitiy of cash desired.
> 9. Customer specifies amount.
> 10. System verifies amount acceptable and provides cash to customer.
> 11. System asks if customer wants to make another transaction.
> 12. Customer selects "No".
> 13. Machine prints receipt and returns customer's card.

The main success scenario describes the successful operation of the system and hence contains directly the key objectives of the system for this particular use. It lists all necessary and expected steps if everything goes as planned.

But things often don't go as planned. Perhaps more important than the main success scenario are the extensions. Extensions are failure modes and alternative paths; anything which could cause a scenario different than the main success scenario. A few extensions to three of the steps in "Withdrawing cash from an ATM" are listed in the table below.

Extensions (Failure Modes)

2. System fails to read card:
    a. Inserted wrong
    b. Invalid card
        i.   Expired
        ii. Withdrawn by bank
        iii. Invalid card (ie, wrong bank)
    c. Card damaged

3. PIN entry failure
    a. User inputs wrong PIN
        i.   Typing Error
        ii. Forgot PIN
        iii. Card Stolen – Theif Guessing
        iv. PIN Changed
    b. Keypad malfunction

10. Cash Amount Error/Problem
    a. More than card limit
    b. More than in bank account
    c. More than in ATM
    d. Not multiple of $20

To develop extensions, consider what can go wrong at each step in the success scenario and/or what other things might occur at each step. Possible failure modes identified via extensions suggest other system objectives.

API RP 75 (SEMS)

The recent SEMS rules make RP 75 mandatory in the GoM. This section identifies specific requirements relative to procedure development and maintenance.

BOEMRE has specified four principle SEMS objectives:

1. Focus attention on the influences that human error and poor organization have on accidents
2. Continuous improvement in the offshore industry's safety and environmental records
3. Encourage the use of performance-based (risk-based) operating practices
4. Collaborate with industry in efforts that promote the public interest of offshore workers safety and environmental protection

API RP 75 applies the following particular requirements relative to operating procedures:

1. Written operating procedures must be in place, designed to enhance efficient, safe and environmentally sound operation
2. Human factors associated with format, content, and intended use should be considered to minimize the likelihood of procedural error
3. Procedure must include the job title and reporting relationship of the person responsible for each of the facility's operating areas.
4. Procedure to cover all important operating scenarios including: startup, normal operations, temporary operations, SIMOPs, emergency shutdown and isolation and normal shutdown.
5. Specify safe operating limits, the consequences of exceeding those limits, and the steps required to correct or avoid a deviation from the limits.
6. Environmental and operational safety and health considerations including:
   a. Special precautions required to prevent environmental damage and personnel exposure (engineering controls and PPE)
   b. Control measures to be taken if physical contact or airborne exposure occurs
   c. Any special or unique hazards
   d. Guidance on allowable/permitted discharge rates/quantities

API RP 75 also requires periodic review of the procedures to ensure that they reflect current and actual operating practices.

CHAPTER 4 – REQUIREMENTS FOR EFFECTIVE PROCEDURES

SOPs serve two basic functions:
1. As instruction to operators on how to operate the facility.
2. As means of checking the design for operability.

Requirements Summary

Simply put, we seek to generate operating procedures that:
- are accurate
- are complete
- contain an appropriate level of detail
- are written early enough to influence the design
- minimize risk including the risk of human error
- are accepted and used by operators
- are easy to modify

Accurate, Complete, Appropriate Level of Detail

Procedures need to accomplish the stated objective when implemented as written. Accuracy and completeness are obvious requirements.

Both of these requirements are somewhat ambiguous; completeness in particular. We could interpret completeness to mean that every required action is included in the procedure. But highly detailed procedures will often be so long and tedious that no one would willingly follow them. Many of the details will be well known to a knowledgeable operator and need not (and should not) be written down.

Here' a trivial example: You want to instruct someone to go to the grocery store to buy some butter. An effective instruction for this task is:
- Buy one (1) pound of salted butter

A significant amount of detail is inferred and need not be specified including:
- Knowing where a grocery store is
- Having a means to get there and back
- Knowing where in the store the butter is kept
- Having a means to pay for the butter and knowing how to check out

How do we determine the appropriate level of detail? A simple approach is to start writing at a High-level (minimum detail) and gradually add detail until additional detail is not useful. (See Chapter 12)

Written early enough to impact the design

SOPs identify operability issues in a design. If the SOPs are developed early enough, the issues identified can be corrected in the design. Early SOPs give us an opportunity to design the facility to be operated rather than trying to figure out how to operate a facility after it has been built.

Minimize the risk of human error

Optimal Complexity

We can consider procedures on a complexity scale as shown below. When we consider options on how to operate a process we may find a very simple method that is relatively risky. Adding some complexity or rigor, some safety checks, warning flags, etc. should make it safer. But at some point we can go too far. When adding complexity makes the procedure hard to implement, the risk will rise.

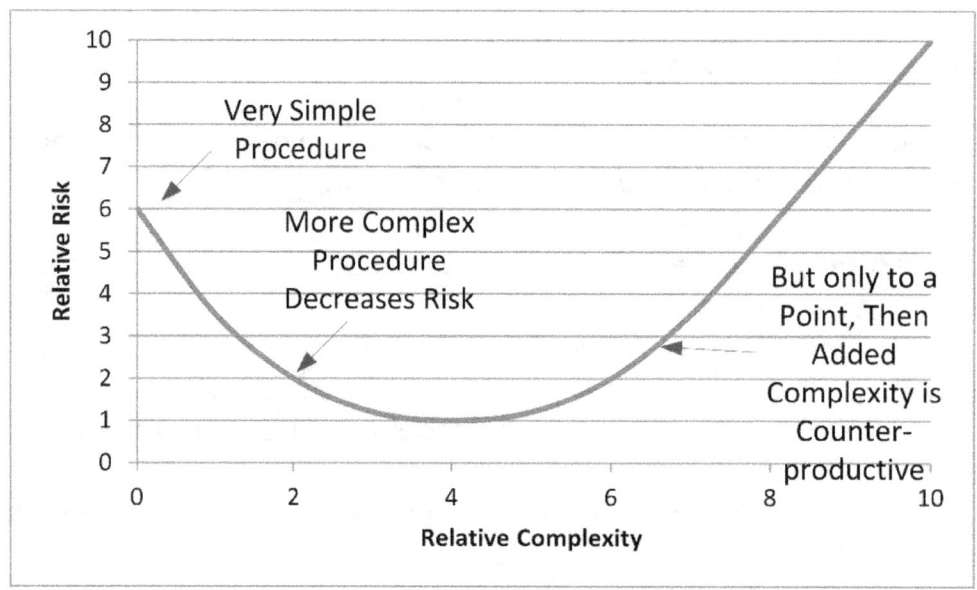

Minimizing Human Error

It is a good start to have accurate and complete SOPs written at an appropriate level of detail, but it is not good enough. Accurate procedures may be implemented incorrectly. We generally call that phenomenon 'operator error'. But some accurate procedures are difficult to follow. It is incumbent on the writer to generate a procedure that minimizes the potential for human error. (See Chapter 11)

<u>Are accepted and used by the operators</u>

All of the above is largely for naught if the operators do not actually use the procedures. Perhaps the best measure of SOP success is to have them accepted by the operators and actually used.

To some extent, this will require culture change as has been accomplished in the airline industry where SOPs, usually in the form of checklists, are used 'religiously'. But this culture change cannot even be attempted if our procedures are not suitable for everyday use.

<u>Easy to modify</u>

SOPs must remain current throughout the life of the facility. Changes will be necessary due to facility modifications. Changes will also be necessary as better ways are found to do things. The SOPs should be written and managed in such a way that making appropriate changes is not overly burdensome.

SECTION 2: THE PROCEDURE WRITING PROCESS

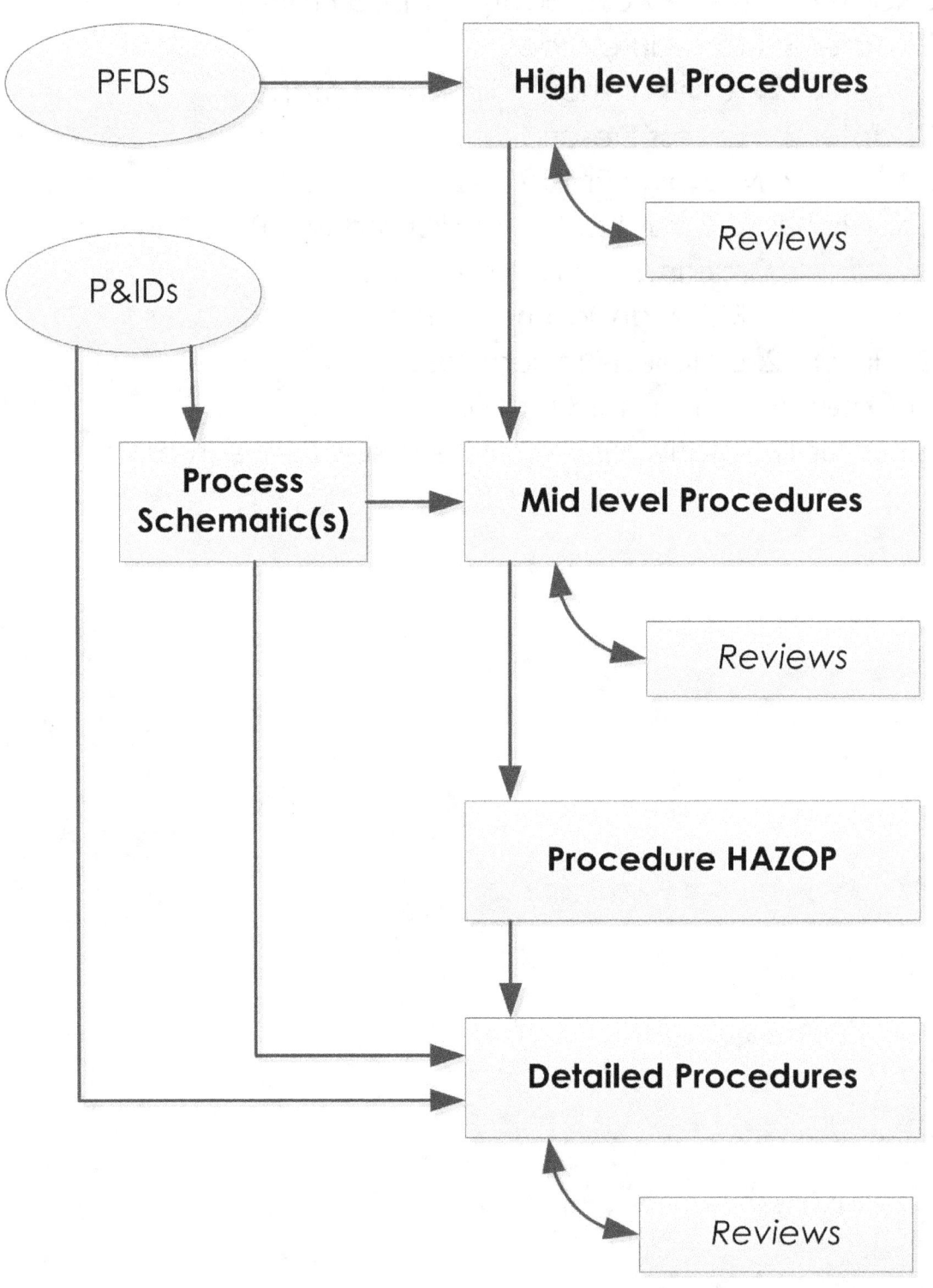

CHAPTER 5 – THE KEY CONCEPT: A PHASED APPROACH

The table below contrasts a procedure containing detailed steps only with a procedure written via a phased approach.

In this example, you see that the detailed steps are all the same – 20 detailed steps in numerical order in each. In fact, if you take a typical 20 step procedure and try to organize it, you will usually find that the steps are not ordered this well; you might have to reorder the steps to achieve this sort of outline structure.

Procedure Composed of Detailed steps Only	Same Procedure with Phased Organization
Detail step 1 Detail step 2 Detail step 3 Detail step 4 Detail step 5 Detail step 6 Detail step 7 Detail step 8 Detail step 9 Detail step 10 Detail step 11 Detail step 12 Detail step 13 Detail step 14 Detail step 15 Detail step 16 Detail step 17 Detail step 18 Detail step 19 Detail step 20	**High-level step I** **Mid-level step I-A** Detail step 1 Detail step 2 Detail step 3 Detail step 4 Detail step 5 **Mid-level step I-B** Detail step 6 Detail step 7 Detail step 8 Detail step 9 **High-level step II** **Mid-level step II-A** Detail step 10 Detail step 11 Detail step 12 **Mid-level step II-B** Detail step 13 Detail step 14 Detail step 15 Detail step 16 **Mid-level step II-C** Detail step 17 Detail step 18 Detail step 19 Detail step 20

There are a number of reasons for writing procedures this way:

1. Procedures written this way conform to the key dictates of structured writing (chunking, limited number of steps to consider at one time, easy to scan). A procedure written this way is easier to read, easier to implement, more likely to be implemented correctly.

2. It is much easier and more efficient to write procedures in a phased approach (write all the High-level steps first, then add the Mid-level steps below each High-level step, then write the detailed steps below each Mid-level step).

3. It is very easy to review the list of High-level steps. It is also easy to review a list of Mid-level steps. Detailed procedures are difficult to write and review. But when the detailed procedures are written specifically to further describe a Mid-level step, those few detailed steps are also fairly easy to write and review.

4. High- and Mid-level Procedures can be written very early in the design and used to identify operability issues in the design.

5. Procedures written this way are much easier to modify and maintain. For example, in the table above if we determine that Mid-level step II-B should follow rather than precede step II-C, then all the details under it follow the change automatically. Similarly, if the order of High-level steps changes, then all subordinate Mid-level and Detail steps follow the High-level steps.

Write in Stages – Don't Just Organize a Finished Procedure

A good deal of benefit is accrued by taking a detailed procedure and organizing it into a High-Mid-Detailed structure. But a great deal more benefit accrues when the procedure is written this way in the first place.

To be perfectly clear on this point, the procedure development follows as shown in the table below.

PHASE 1	PHASE 2	PHASE 3
Write and Review the High-level steps First	*After the High-level steps are Agreed, Write and Review the Mid-level steps required to define each High-level step*	*After the Mid-level steps are agreed, and the design is Sufficiently Matured, Write and Review the Detailed steps under each Mid-level step*

High-level step I
High-level step II
Etc.
	High-level step I
	   Mid-level step I-A
	   Mid-level step I-B
	High-level step II
	   Mid-level step II-A
	   Mid-level step II-B
	   Mid-level step II-C
	Etc.
		High-level step I
		   Mid-level step I-A
		      Detail step 1
		      Detail step 2
		      Detail step 3
		      Detail step 4
		      Detail step 5
		   Mid-level step I-B
		      Detail step 6
		      Detail step 7
		      Detail step 8
		      Detail step 9
		High-level step II
		   Mid-level step II-A
		      Detail step 10
		      Detail step 11
		      Detail step 12
		   Mid-level step II-B
		      Detail step 13
		      Detail step 14
		      Detail step 15
		      Detail step 16
		   Mid-level step II-C
		      Detail step 17
		      Detail step 18
		      Detail step 19
		      Detail step 20

Easy to write.
Easy to review.
Easy to scan.
Easy to change.
	Easy to write.
	Easy to review.
	Easy to scan.
	Easy to change.
		Detailed steps under any given Mid-level step are:
		   Easy to write.
		   Easy to review.
		   Easy to scan.
		   Easy to change.

CHAPTER 6 – PROCEDURE MAPS

Procedures should be short lists of action steps to be performed in chronological order.

Where a large number of steps are required the procedure should be divided.

Where significant decision making and potential branching is required the decision making and branching should be captured outside of the procedures themselves.

Procedures Should be Short

Procedures should generally be short and consist of a list of steps to be performed in chronological order. But a procedure to start up an oil and gas facility will likely be fairly long.   This is generally mitigated by writing procedures for individual systems rather than writing one large procedure for the entire facility. Procedure maps provide a graphic method of ordering the individual procedures.

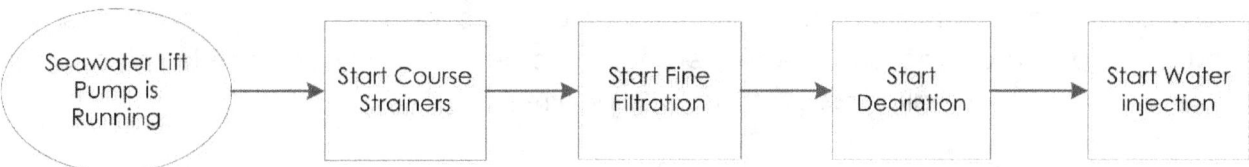

This is a procedure map for a waterflood system startup. The waterflood startup procedure could be written as one long procedure, but it would contain dozens of steps. The schematic serves to divide the procedure into manageable chunks. Each block on the schematic can be written as a separate procedure or each block can be a High-level step.

Avoiding Branching and Decision Points

In general a procedure should be a list of steps to be performed in chronological order. But in a complex system there may be decision points that cause branching.

It is possible to include decision points in a procedure. These would generally be of the form:

8	If Condition = A then go to step 9
	If Condition ≠ A then go to step 11
9	Action(s) required if Condition = A
10	.....
11	Action(s) required if Condition ≠ A

This sort of logic should be kept to a minimum. It makes procedures more difficult to follow and also makes them more difficult to change and maintain. In general we should avoid referencing procedure step numbers because any change in the number of previous steps will make these instructions incorrect.

A more effective strategy is to include the requisite instruction within the decision step:

8	If Condition = A then:
	Action(s) required if Condition = A
9	If Condition ≠ A then:
	Action(s) required if Condition ≠ A
10	.....

This works if the alternative instructions are simple.

When the branching involves significant changes, then the decision making and branching should be captured in one of two ways:
1. via reference to another procedure or
2. via a procedure map.

Branching via Reference to a Separate Procedure

8	If Condition = A then proceed to next step.
	If Condition ≠ A then:
	Apply SOP 16
9	…..
10	…..

Procedure Maps with Logic

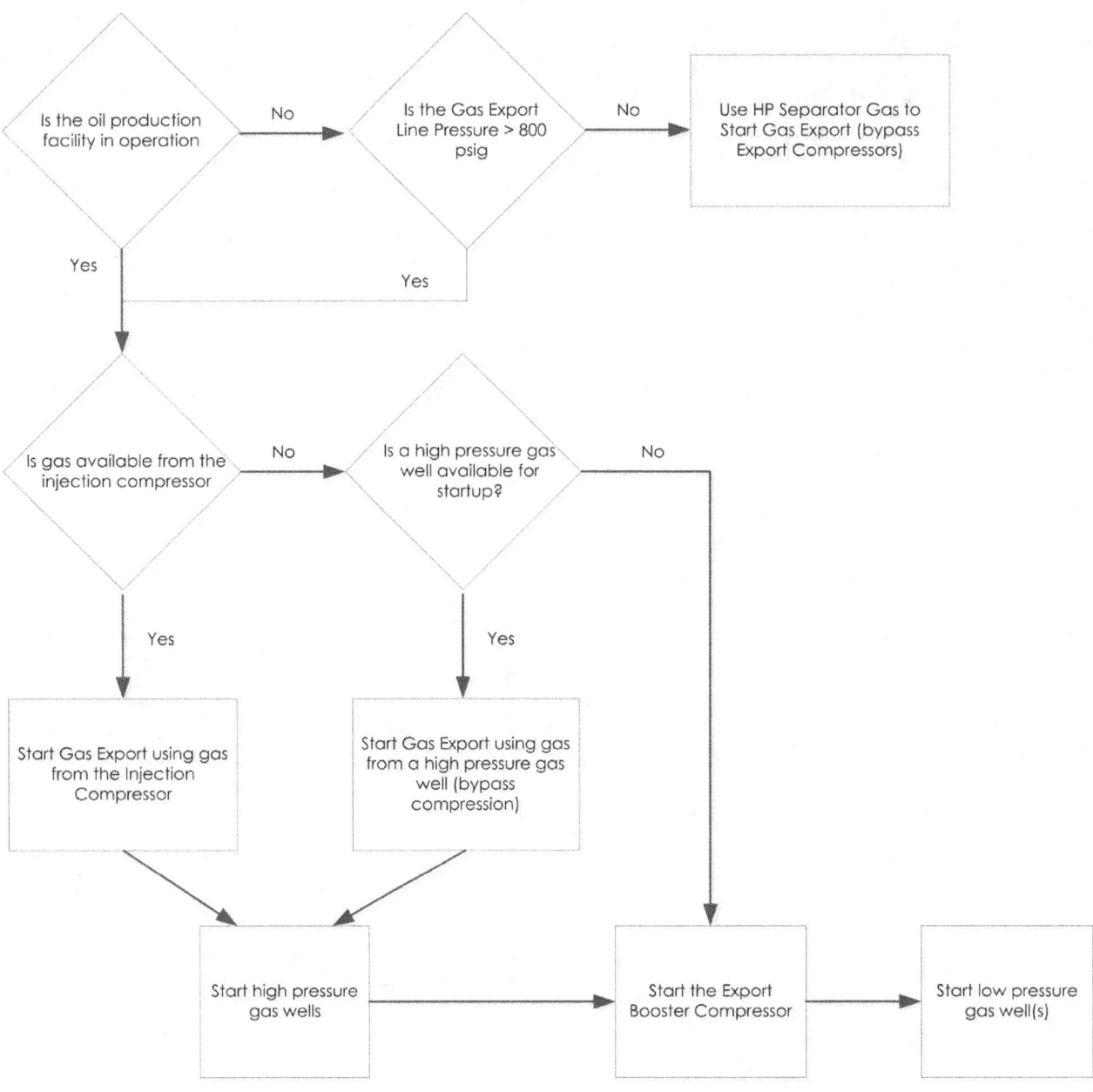

This procedure map is for the startup of a gas export facility that can be started in different ways depending on sources of gas available and the current operating pressure of the Gas Export Line. Each startup sequence is different depending on the decisions made. But the decisions are all made at the procedure map level and the individual procedures (rectangles) will not contain decision points.

CHAPTER 7 – PROCESS SCHEMATICS AND SKETCHES

The definitive reference documents for process procedures are the P&IDs.

But P&IDs are too detailed for most operating discussions. A single procedure may involve many P&IDs.

Using the P&IDs is cumbersome, creates tunnel vision, and tends to drive engineers to write the details too early. And, of course, the P&IDs cannot be readily copied into the procedure. Simpler drawings are required both for reference during procedure development and for inclusion in the procedures themselves.

Overall Process Schematic

Our solution is to generate a Process Schematic showing the entire process to be operated on one sheet. The sketch below is one we drew for a gas production facility startup. This schematic summarized over 40 P&IDs. Once we finished it, we put the P&IDs away until we needed them again for writing the detailed procedures.

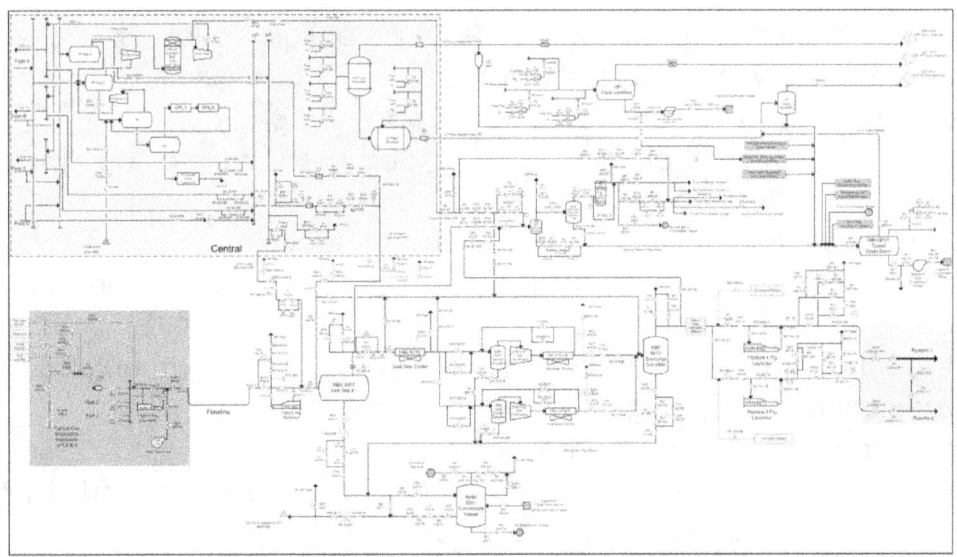

Standard for Overall Process Schematics

The process schematics summarize the P&IDs. The entire process should be included on a single sheet to be displayed and plotted as a single drawing. If this cannot be reasonably accomplished, a block diagram may be drawn showing the entire process on a single sheet with individual systems on other sheets. Our standard for overall schematic drawings is:

- All actuated valves are shown
- Manual chokes that are adjusted to control flow are shown
- Manual valves are shown only when they are important to the startup and are called out in the procedure. Manual valves addressed only in valve lists are not included.
- Check valves are shown
- Drains and vents are not shown typically. Exceptions are vent and drain valves specifically called out in the procedures.
- All control loops are shown. Where the loop design is obvious, it is acceptable to show only the control valve. Complex loops are shown in enough detail to describe their operation.
- The failure positions of control valves are shown.
- No vessel internals are shown except potentially buckets, weirs and trays necessary to show the path of fluid phases.
- PSVs are not shown (they have no active role in operation).
- Safeguarding switches are shown. It is also useful to provide a table of their actions (summarized cause and effect chart)
- P&ID boundaries are shown to permit easy reference.
- Equipment items are shown as small as practical to save room for piping and valves.
- Equipment capacity, size and design ratings may be shown, but are typically not as that clutters the drawing.
- Variable speed drivers are clearly indicated as such.
- Where pumps or other equipment items are started automatically by the control system that should be indicated via notes.
- Utilities are shown simply on the overall process schematic and may be shown in more detail on utility schematics if required. For example, heating media supply to users is shown simply as an HMS arrow. The entire Heating Medium System is shown on a separate schematic.
- Optional – summarize cause and effects

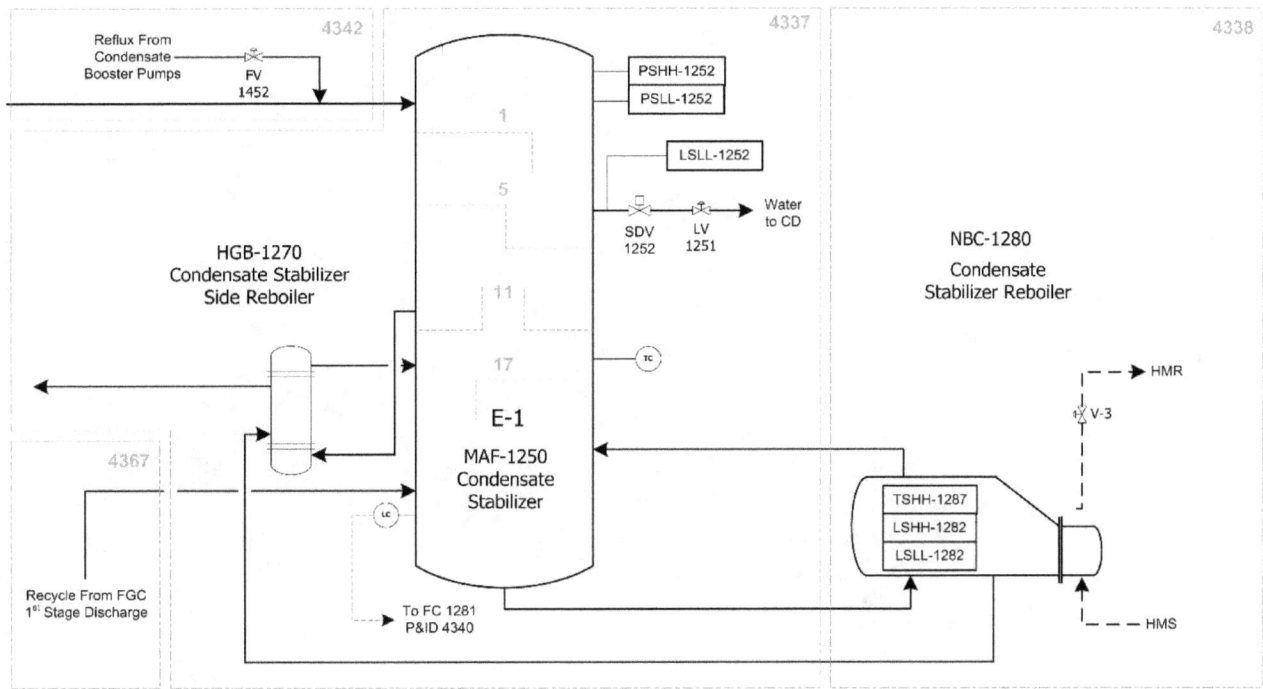

Example Schematic Section to Show Features and Level of Detail

CHAPTER 8 – PROCESS DESCRIPTION

At the risk of stating the obvious, both the procedure writer and the procedure implementer (operator) must be very familiar with the process.

A process description should be included in the procedure document or in a separate document readily available to operators.

If the project has a process description and/or a control narrative, use that as the starting point and edit as necessary per the rules of structured writing to make the information more accessible. Include in the process description everything that an operator needs to know to successfully operate the system.

Include information relevant for day to day operation without excessive detail.

A picture is worth a thousand words. Use sketches liberally for describing the process.

Example

See the procedure in the Appendix for an example.

CHAPTER 9 – MID-LEVEL PROCEDURES

Many Mid-level Procedures follow a similar pattern. The purpose of most procedures in the processing industries is to establish, stop, or change a flow. Hence, the key issue will usually be either to control the flow with control valve or with a piece of rotating equipment. A prototypical list of Mid-level steps for a process system startup is:

A. Preliminary Checks and Actions

This will include actions that can and should be performed early (off critical path) before the system is ready for startup. Examples are checking the lubrication in rotating equipment items, setting important boundary isolations, setting selected shutdowns in bypass or startup condition, confirming that assumptions and pre-requisites are satisfied.

B. Set Control System for Startup

If the procedure establishes or changes the flowrate of a stream, the control point should be identified and set early in the procedure. Often this step will close the key control valve to ensure that there is no flow before the system is ready.

C. Align Manual Valves

Many of the manual valves may be set in step A, but some may have to wait until the control system is set.

D. Bypass/Inhibit Safeguarding System as Required for Startup

Ideally there will be a startup mode in the safeguarding system that inhibits safety instruments as necessary for startup (for example – inhibiting a low pressure shutdown until the system is at operating pressure). Any instruments that must be manually bypassed and then manually reset later should be explicitly called out.

E. Establish Flow

The key control valve or choke is gradually opened to establish flow. And/or a piece of rotating equipment is started. Any required chemical injection will likely be started here.

F. Monitor Operating Conditions

As the flow is increased, process conditions in the system will change in predictable ways. These expected changes should be explicitly described and monitored. This may be the most important part of the procedure. We want to catch errors and equipment malfunctions as early as possible.

G. Ramp to Target Rates/Conditions

In steps E and F we establish flow, often at a minimum rate, and then confirm that the system is operating as expected. Here we ramp the flow (or temperature, pressure, level) to the target operating condition.

H. Reset Safeguarding System for Normal Operation

Manually reset safeguarding systems as necessary.

I. Completion steps

Capture here anything that needs to be done after the startup is 'complete'. This may include stopping injection of chemicals only needed during startup, removing personnel barricades erected for startup, completing paperwork, etc.

CHAPTER 10 - OVERALL STARTUP AND SHUTDOWN PROCEDURES VIA ORDERING OF MID-LEVEL STEPS

Procedures are generally written for individual systems, but during a platform startup many systems are started more-or-less simultaneously. How can we coordinate the action steps from multiple system procedures into a single overall startup procedure?

Consider the prototypical system startup procedure given in chapter 9:
- A. Preliminary Checks and Actions
- B. Set Control System for Startup
- C. Align Manual Valves
- D. Bypass/Inhibit Safeguarding System as Required for Startup
- E. Establish Flow
- F. Monitor Operating Conditions
- G. Ramp to Target Rates/Conditions
- H. Reset Safeguarding System for Normal Operation
- I. Completion steps

Now consider startup of the five systems as shown on the procedure map below:

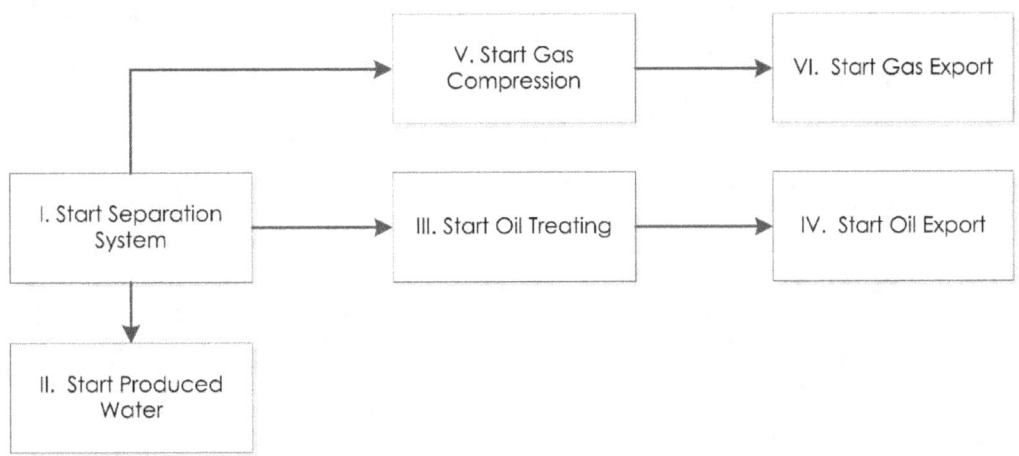

An overall startup sequence can be built by ordering the Mid-level steps as shown on the table below.

Time Period	System I	System II	System III	System IV	System V	System VI
1	A B C D	A B C D	A B C D	A B C D	A B C D	A B C D
2	E F					
3		E F	E F			
4				E F		
5					E F	E F
6	G	G	G	G	G	G
7	H I	H I	H I	H I	H I	H I

CHAPTER 11 – RISK MANAGEMENT - PROCEDURE HAZOP

In Chapter 4 we noted that there is, conceptually at least, an optimal level of complexity at which risk is minimized.

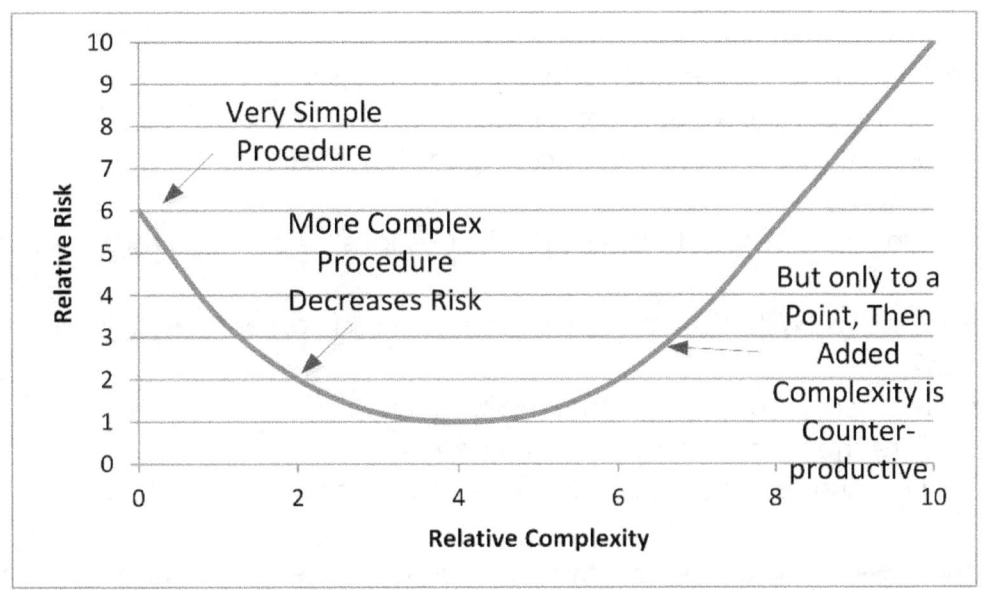

SOPs are written mainly to minimize operating risks. Risks can be categorized as:
1. Inherent process hazards
2. Risk of operator error

Both must be addressed. Procedures must be written to minimize risk when implemented correctly. But that is not sufficient. They must also be written to minimize the probability of human error.

We take the following approach to procedure risk management:
1. The procedure writer should explicitly identify inherent hazards prior to writing the procedure.
2. At each step in the process the writer/team should ask "What could go wrong at this step?"
3. An effective procedure HAZOP should be performed on the Mid-level Procedures.
4. Management systems and operating culture must be in place to ensure appropriate use of the procedures in the field. This may involve categorizing procedures according to risk level.

The procedure HAZOP is performed on the Mid level procedures. These procedures have an appropriate level of detail for performing a HAZOP; enough detail to identify significant risks, but not so much that the study team gets lost in the weeds.

PROCEDURE HAZOP METHOD

The review method is fundamentally a two-step process:
1. Describe the planned operation of the system
2. Brainstorm what could go wrong during that operation

The brainstorming is structured via a set of questions. For the overall procedure we consider:

- Effectiveness (Will it accomplish the desired objective if implemented as written?)
- Completeness (Are any steps missing?)
- Order (Are the steps listed in the right order?)
- Impact of missing, incomplete, incorrect assumptions or pre-conditions

Following that, we evaluate each Mid-level step in the procedure with the aim of answering the question "What could go wrong at this step?" The discussion is structured via the guidewords listed below:

Table 1: Questions for Individual steps

Equipment
- What if equipment item fails?
- What if equipment is in unexpected state?

Process Limits
- Pressure: Can the system Pressure limit be exceeded?
- Temperature: Can system temperature limits be exceeded?
- Level: What if level limits are exceeded?
- Flow: Is flow induced damage possible? (vibration, erosion, severe slugging)

Human Error
- Are the requirements clear or ambiguous?
- Is the step obvious/common or unusual?
- Is the step simple or complex?
- Does the step extend over a long period of time?
- What if the step is skipped? Is there latent error potential? (skipped step won't be noticed)
- What if the operator does too much or too little?

- What if the operator does this step too early or too late or out of sequence?
- What if the operator does something different?

Other Issues?

METHOD BASIS

The selected study method is based on the
1. Human error research (James Reason, 1990)
2. Observations from cognitive science research (Fiske and Taylor, 1991)
3. Typical Procedure HAZOP methodology

Cognitive science research identified this important fact: we are pretty good at spotting errors in a procedure, but relatively poor at spotting missing steps or omissions. Hence an early step in this Procedure HAZOP is inspection for omissions.

Reason has catalogued many causes of human error. The important ones for procedure reviews are summarized below:

	Pre-conditions are likely to be incomplete or not done if or when:	
1	Following maintenance/repair the status of equipment may not be as expected. *For instance: slip blinds left in place, equipment not powered, safety systems bypassed, instruments on manual, etc.*
2	Checklist does not follow logical progression. *For instance: valves in a line required to be in a given position should be listed in line order to facilitate checking.*
3	Required conditions prior to proceeding to next step are ambiguous
	Errors in implementing steps in a procedure are most likely:	
4	If a step is not obviously cued by the previous step it is likely to be skipped
5	If there are too many steps, steps in the middle are likely to be skipped
6	If a single step is complicated and has sub-steps, slips and lapses are likely
7	A step which extends over long period of time without intervening action is prone to distraction. The operator may not notice cues to continue to the next step. And the operator may forget where he is in the process and may skip steps or repeat previously accomplished steps.
8	Steps which occur after the main goal is achieved are likely to be skipped.
9	Interruptions from SIMOP's (simultaneous operations) can cause you to lose your place.
	Latent Errors – Problems waiting to happen	
10	Look for latent errors caused by skipped steps. If a step can be skipped without impacting further steps or causing immediate upset to the process, then it is likely that failure to perform the step will not be caught. Somehow reinforce the need to complete the step.

PROCEDURE HAZOP DOCUMENTATION FORM

Procedure: _____

OVERALL PROCEDURE EVALUATION

Effectiveness (Will the procedure as written achieve the desired objectives?)

Completeness (Are any steps or prerequisites missing?)

Order (Are the steps in the right order?)

What if an assumption is incorrect? What if a condition precedent is incorrect or not achieved?

EVALUATION OF INDIVIDUAL STEPS (consider each topic for each applicable step in the procedure)

	Discussion/Recommendations
Equipment • Failure • Status	
Process Limits • Pressure • Temperature • Level • Flow • Erosion	
Human Error • Clear/Ambiguous • Obvious/Unusual • Simple/Complex • Long duration • Latent errors • Too much/little • Too Early/late • Different	
Other	

CHAPTER 12 – DETAILED PROCEDURES

Detailed procedures are written only after the Mid-level procedures are reviewed and accepted/approved. They are only written to further define the Mid-level steps. Not every Mid-level step will have to be detailed further. Additional detail should be added where needed, but only where needed.

Types of Details

Several types of details can be included:

1. Action command

    This is the most common type of detail. It is simply an instruction to do something such as 'Open valve 101'

2. Monitor command (expectations)

    Operators should have an expectation of what will happen when an action is taken. Sometimes these expectations are important enough to include in the procedure. Sometimes the next step cannot be taken until an expected condition is reached.

    Example: Pressure in the Production Separator will increase to 1350 psig at which point PIC-104 will begin controlling pressure.

3. Operating Limits and Set Points

    Operating limits and critical set point with safety implications should be included. Operating limits are required by SEMS in the GoM.

4. Checklist (as per CRM)

    Chapter 3 summarizes the use of checklists as they are used in the airline and medical industries. In both cases, the checklists are not fully detailed procedures; they are lists of things that a competent operator might forget. Such checklist items are very appropriate in a detailed section.

5. Explanatory notes or diagrams

If a Mid-level step is ambiguous, a short note or sketch may be sufficient to clarify the intent.

6. Warnings

Where a hazard exists, it should be identified in a standard way (bold, different background color, icon, etc.)

The procedure format should distinguish between each type of detail.

How Much Detail is Required?

There is no simple answer to that question. Enough detail must be provided to ensure that the procedure will result in effective operation, but not so much that the operator will be overwhelmed with detail.

Note that the GATE process makes it easier to agree on the appropriate level of detail for any given procedure. A discussion is had during the review of the Mid-level procedures. Details are only added when needed.

The required level of detail for a given procedure is a function of:

1. The experience and knowledge level of the operator

A novice may need a more detailed procedure than a senior operator. Procedures are generally written for the control room operator who is (tacitly at least) assumed to have a fairly high skill/knowledge level.

2. The complexity of the task / frequency of use

Tasks that are intrinsically easy or which the operators are very proficient at because they do them often, in general should not require as much detail as difficult and infrequent tasks.

Different types of complexity need to be considered.
- Some tasks are difficult simply because of the number of steps that need to be done correctly and in the right order. In that case a detailed, step-by-step procedure is certainly in order.
- Some tasks require more intimate knowledge of the process dynamics and require judgment or 'touch'. In this case the appropriate

approach may be to limit the actions to experienced staff rather than writing a complex procedure.

3. The importance of the task

    In general, the more important the task (safety, environmental, economic) the more detailed and explicit a procedure should be.

4. Ambiguity or Uncertainty in the Mid-level step

    Mid-level steps are short and concise. As a result, they may be misinterpreted. Effort must be put into making the Mid-level steps unambiguous. Detailed procedures are certainly one way to do that.

    Even if the goal of a Mid-level step is clear (Start Pump A) there may be multiple ways to accomplish it. If there is a particular way that the step should be performed, then details can convey that.

Scannable

One of the important concepts from the field of structured writing is that a document should be scannable. Whatever level of detail you choose to include in a procedure it is critical that the procedure remain easy to reference and use.

Mid-level steps make the document scannable. To maintain this, the number of detailed steps should be kept reasonably low. Nine (7+2) steps is a reasonable limit. If there are more steps than that, consider whether the Mid-level step can be split into two steps.

Format

A standard format should be applied to all procedures in a set. We consider the following aspects to be important for judging whether a given format is adequate:
1. The procedure should contain both instructions and comments and it should be clear which is which.
2. The more critical steps and conditions should be readily apparent via color coding, bold text, etc.
3. Graphics and data tables or plots should be integrated into the procedure or otherwise made readily available.

Below is an example of an effective format. Note the following:
- Mid-level steps are bold providing easy scanning.
- White space above and below each line of text make reading easier.
- You will notice that this example is printed in grayscale, but if printed in color:
  - Caution statement has yellow background and black text with more 'white' space.
  - Critical safety information has a red background and white text.
- Notes are in smaller font and italics.
- Borders are a shade of grey. They provide delineation, but don't detract from the words.

No.	Action	Tick
A. Start Seawater Jockey Pump (P&ID-4615)		
1	START Seawater Jockey Pump No. 1 PBE-7060 from Hand Switch HS-7060.	
2	SLOWLY OPEN PCV-7061 to slowly fill and pressurize the downstream equipment.	
3	START Copper Ion Generator ZZZ-7050 and start flow to Jockey Pump.	
B. Start 1st Seawater Lift Pump (P&ID-4556)		
Caution: Water Hammer Potential on pump startup. See next two steps.		
4	Confirm that the downstream equipment is water-filled. *The coarse strainers are filled using a Jockey Pump via the procedure above*	
5	Confirm that the overboard valve is properly set and aligned for operation. *Water will flow overboard initially to dampen water hammer effect of pump startup*	
6	START Seawater Lift Pump No. 1 PBE-7010 from Hand Switch HS-7010.	
HAZARD: It is critical to avoid ...		
7	….	

Embedded Graphics

Simple schematics help operators understand the procedure. Where practical they should be embedded in the procedure.

No.	Action	Tick
1	Align Hot Oil from the Hot Oil Storage Tank through the motor driven Hot Oil Pump, PBE-6250. 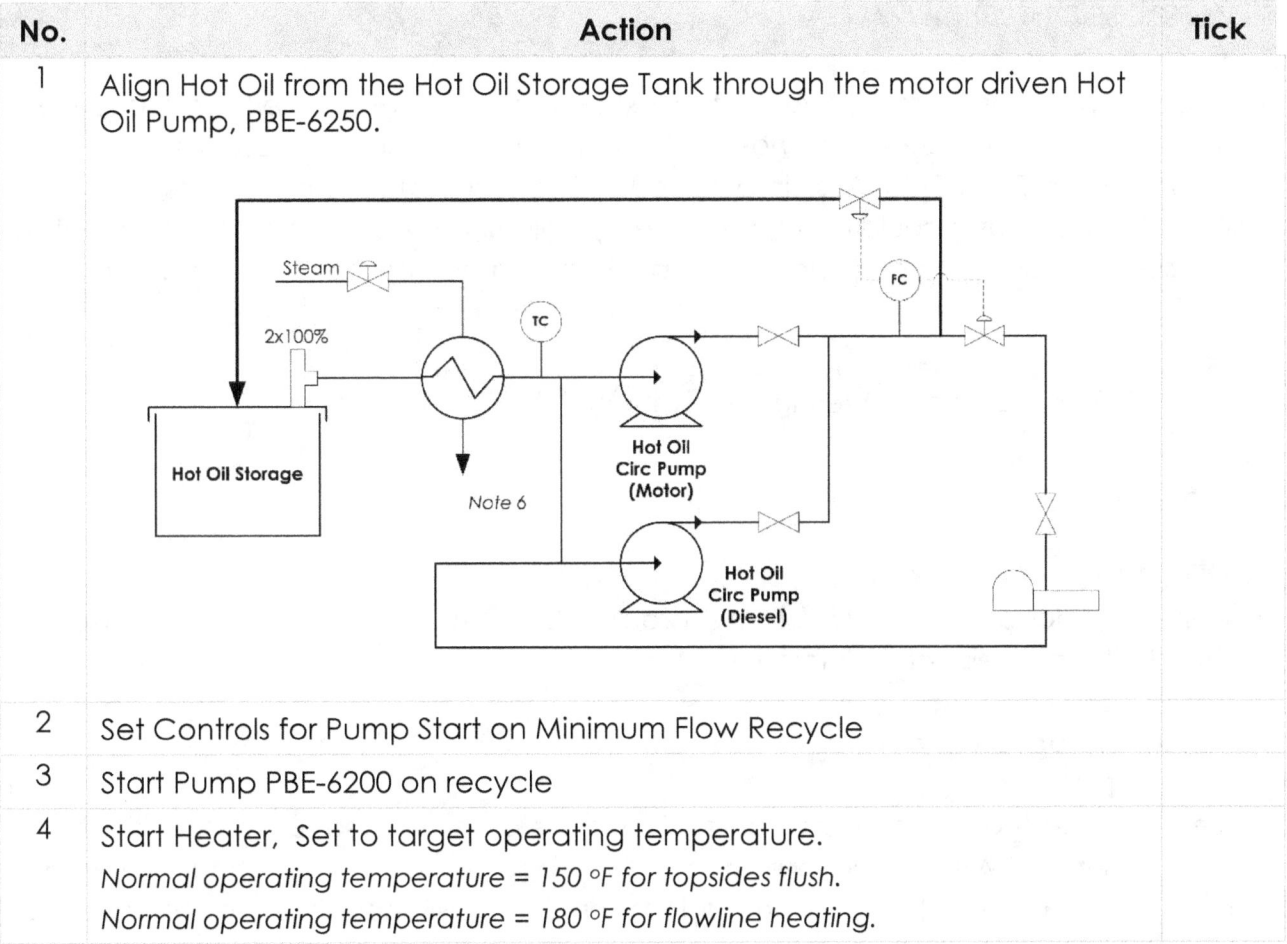	
2	Set Controls for Pump Start on Minimum Flow Recycle	
3	Start Pump PBE-6200 on recycle	
4	Start Heater, Set to target operating temperature. *Normal operating temperature = 150 °F for topsides flush.* *Normal operating temperature = 180 °F for flowline heating.*	

Language

Action steps should have similar language. There are many ways to say the same thing, but some might be interpreted differently or be difficult to interpret. By standardizing language we increase the probability that instructions will be interpreted correctly.

Here are a few ways to instruct an operator to close a control valve:

- Close PV-1251
- Close PV-1251 on manual
- Set PV-1251 to manual closed
- Set PIC-1251 on manual and close PV-1251
- Put PIC-1251 in manual and set output to 0

- Put PIC-1251 in manual and set output to zero

From this list our preference is number 2 "Close PV-1251 on manual" though actually with 'close' and 'manual' in caps. This is short and clear.

> CLOSE PV-1251 on MANUAL

We occasionally see wording such as the last two lines. They are utterly unacceptable as it is not abundantly clear that those commands are intended to close the valve. The first item "Close PV-1251" is the simplest, but it is ambiguous. Since it doesn't specify 'manual' the operator could close the valve by adjusting the set point. If the objective is to close the valve by changing the set point then a command such as the following should be used:

> CLOSE PV-1251 by lowering the set point to 5 PSIG below the current pressure.

Ambiguity

In general procedure steps should be precise, but some level of ambiguity is often required. The quintessential ambiguous step is:

'Crack open valve ---'

While we agree that more precise phrasing is preferable, we also realize that this is not always possible. When such ambiguous language is used, there should be some direction for judging when the amount of 'cracking' is enough or too much. For instance:

> Crack open valve V-101 to establish flow through Filter, F-100
> *At a low flowrate the pressure drop through the filter should not exceed 2 PSI.*

CHAPTER 13 – ABNORMAL SITUATIONS

Some abnormal situations will require procedures.

Abnormal operating situations may include:
1. Operating with impaired or out-of-service process equipment items
2. Operating with impaired or out-of-service utility systems
3. Operating with impaired control or safeguarding systems
4. Operating outside of design operating conditions or outside of a specified operating window
5. Operating with limited staffing
6. Operating with simultaneous other operations ongoing (SIMOPs)

It will not be possible to identify every possible operating situation, nor is it necessary or wise to write procedures for each identified situation.

It is necessary to identify the situations in which the plant will operate in an impacted state and to appropriately address those with procedures or operating instructions.

Suggested approach:
1. Identify equipment items that can be isolated for maintenance. Write procedures for operation without the items if necessary.
2. Review the prerequisite list. Is operation possible with one or more of the prerequisites not satisfied? Do you need a procedure for such operation?
3. Review the assumptions list. What if one of them is not satisfied? Is a procedure needed?
4. Does the procedure identify required staffing? What if a shift is short-handed?
5. Evaluate control valves for criticality. Which ones can you bypass (control the process manually)?
6. Identify possible SIMOPs. Which ones can you operate with? Is a procedure required?

CHAPTER 14 – INTERFACE TO MAINTENANCE PROCEDURES/TASKS

A set of maintenance procedures (repair and preventive) should exist. There is an interface between actions performed by operators and actions performed by maintenance technicians.

Some tasks are performed exclusively by maintenance, with operations actions to isolate the equipment.

Some tasks may be the responsibility of either operations or maintenance depending on platform operating philosophy and staffing. This may involve tasks such as cleaning, lubricating, testing/checking, etc.

Example: Replacing Filter Elements

Consider a task of changing filter cartridges. Whether the task itself is assigned to operations or maintenance, there are aspects of the task that are operations-centric and aspects that are maintenance-centric.

Operations-centric tasks:
- Identify need for new filter cartridges (via pressure drop measurement)
- Isolate the filter housing from the process
- Drain, vent, flush, etc.
- Return filter to service after element replacement

Maintenance-centric tasks:
- Identify parts needed (number and type of elements, gaskets, etc.)
- Obtain parts from stores
- Reorder parts as necessary (maintain inventory)
- Remove and replace the filters elements
- Dispose of the old elements.

The procedure to replace the elements will not normally be included in the SOPs. Some plan should be in place to capture these types of procedures and an appropriate interface provided to the SOPs and the maintenance planning system.

Example: Firewater Pump Testing

Another good example of an interface requirement is firewater pump testing. Firewater System SOPs will include directions for normal operation. But most of the operations of the firewater pumps will be test runs; generally weekly test runs and annual test runs at least.

Procedures for these tests may be included in the SOPs or in the maintenance procedures or both. Actual implementation of the test runs will likely be a joint effort.

SECTION 3: PROCEDURE REVIEWS AND OPERATOR TRAINING

Operators need a variety of knowledge and skills. With respect to the SOPs operators must:

- Have the operating skills, knowledge, and experience required to understand and implement procedures in general.
- Have sufficient specific knowledge of the facility to be operated in order to understand and implement the procedures for that facility.
- Understand the facility process well enough to be able to respond appropriately to abnormal situations.

For the purposes of this chapter we assume that the operators to be trained are competent operators who need training on:

1. The facility process design. A fairly detailed knowledge of the process is required to be able to operate it.
2. The facility SOPs.

We expect the operators to be able to implement the procedures as written. But we hope for something more –

> *We hope that an operator will have the knowledge and confidence to respond effectively to abnormal events not covered in the procedures.*

Two training goals are appropriate:

1. Teaching operators the content of the procedures and how to implement them.
2. Teaching them what to do if something goes wrong.

The procedure review process is included in this chapter because a procedure review is a good opportunity/method for training. The GATE process requires procedure reviews at all three levels. Each of these reviews is an opportunity for operator training.

CHAPTER 15 – PROCEDURE REVIEWS

High-level Procedure Reviews

High-level procedures are short and easy to review. These reviews are usually done via unstructured, ad hoc methods.

Mid-level Procedure Reviews

Mid-level procedures are also relatively short and are easy to review.

After the Mid-level procedures are written for all systems, we order the Mid-level steps into an overall process startup procedures. That effort frequently identifies errors and omissions in the Mid-level procedures.

The Procedure HAZOP is done at this level and provides a very effective review.

Another very effective method is to 'implement' the procedures on the overall process schematic. In this process we walk through the procedure step-by-step, marking up the schematic as we go (open valves green, closed valves pink, energized pipe and equipment yellow).

Detailed Procedure Review

This review is frequently performed as an operator training exercise. We pin or tape the P&IDs to a large conference room wall and mark them up according to procedure steps (open valves green, closed valves pink, lines with flow/pressure yellow).

CHAPTER 16 – OPERATOR TRAINING MANUAL

A great deal of expertise goes into the writing of SOPs; they are typically written by senior operators and or senior process engineers with assistance from other subject matter experts. The resulting procedures are chronological lists of action steps plus some of the most important background information.

The expertise that went into writing the procedure cannot be included in the procedure. Procedures must be short and accessible for use 'in the heat of the battle'. The operators must possess enough process expertise to understand why the procedures are written the way they are – and by extension, to be able to effectively deviate from a procedure when necessary.

To keep the procedures simple and useable, the additional information that operators need for training and reference should be included in a separate document (Standard Operating Manual or Training Manual). The training manual should contain:

1. A process description
2. Equipment descriptions
3. A detailed discussion of the SOPs. (Why they are written the way they are. What opinions were considered and rejected.)
4. Questionnaires to prove understanding.

CHAPTER 17 – SCENARIO –BASED TRAINING

Abnormal Situation Management (ASM) describes an abnormal situation as "a disturbance… in a process with which the control system is unable to cope and which required operator intervention."

An important goal of operator training should be to equip them to respond appropriately to abnormal situations.

Some abnormal situation can be anticipated and procedures written to manage them. In this case, typical procedure training may be effective. But since the abnormal situation occurs infrequently, that training may have to be repeated periodically to be effective when needed.

Some abnormal situations will not be anticipated and/or no procedures will be written. How can we train operators to respond effectively to novel upsets? The field of Naturalistic Decision Making (chapter 3) provide some insight. To respond effectively to novel events operators must possess:
- Pattern recognition skills (ability to recognize an event early from the first cues that something is going wrong)
- Mental simulation skills (ability to predict what will happen next if nothing is done, and to simulate what will happen if the operator takes particular actions).

Scenario-based training methods build such skills.

HAZOPs as a Training Vehicle

A novel approach to operator training is to perform a HAZOP with them. We want operators to thoroughly understand the process and the operating procedures.

The HAZOP process it typically used to determine how the plant control systems and safeguarding systems should respond to upsets. But we are interested in teaching CROs how **they** should respond to upsets; so, a modified HAZOP process is used. The first three steps are the same as a typical HAZOP process:
1) Select Applicable Guideword-Deviation Pair
2) Ask if one or more causes for [deviation] [guideword] exists
3) If at least one cause exists, ask if the deviation has significant consequences

From there we diverge. The remaining steps are:
4) How will you recognize this problem?

5) What will happen if you take no action?
6) What response should you take to solve the problem?
7) What impact will that response have on the process?

This process identifies upset scenarios and has the operators work through responses. This process can be used also to identify scenarios to be used for scenario games as described below.

It is not useful to consider all usual HAZOP guidewords in this process. Changes in flow cause changes in pressure, temperature and level. Use only the 'flow' guideword and draw larger nodes.

Role of Instructor

The CRO training HAZOPs are largely CRO driven. The role of the instructor in the HAZOP exercise is to facilitate, answer questions and correct errors. The CROs do most of the talking. This has the beneficial effects of reducing tedium (as compared to lecture type training), and effectively identifying gaps and errors in CROs knowledge. A lecturer never really knows if his students understand. In this program, the students are doing all the talking and the instructor listens for errors.

Inspiration for the HAZOP-based training program came from an article by H.C. Sayre (Sayre, 1994).

Operator Training via Scenario Games

The HAZOP-based training described above is effective in part because it identifies scenarios that operators work through. In that situation, the operators are generating their own scenarios via the HAZOP process. Scenarios can be developed by others and used directly for training.

Tactical Decision Games (TDM)

The US military, police organizations, firefighting teams and other groups in dynamic businesses use a process called TDM for training.

The TDM process is simple:
1. Identify a situation. Typically the situation will be dynamic (changing fast) and not completely defined (ambiguous cues, conflicting cues, etc.)
2. Give the operator a very short period of time to react to the situation where the required action may be to list as many possible causes as the operator can think

of, to identify the worst possible consequence, to identify who should be alerted and when, to suggest what data should be collected to clarify the situation, etc.

Example: Dropping Separator Level?

The production separator has level has been stable. There has been no obvious system changes that would affect the level. Indeed the control system level transmitter is showing steady level at the normal liquid level and the level control valve has not moved recently.

But the safeguarding transmitter is showing a rapidly dropping level and it will soon hit the level switch low and shutdown the plant.

Instruction:
1. Immediately give an instruction to the outside operator to: _____.
2. Next: You have 2 minutes to identify all possible causes for the discrepancy between the two level transmitters.

CHAPTER 18 – OPERATOR TRAINING SIMULATORS

Operator Training Simulators (OTS) are useful to a project in several ways including:
1) Effective means of testing the control system
2) Effective means of testing the SOPs
3) Means of implementing scenarios

Testing the SOPs

If the OTS is of high enough fidelity, it will be possible to implement the SOPs step-by-step. This is a very good test of the SOPs themselves and provides effective training for the operators on implementing the SOPs.

Implementing Scenarios

OTSs are particularly valuable for modeling abnormal situations. A particularly instructive method is to play scenario games, as described in Chapter 16, and then run the scenarios on the training simulator to see if the operator's mental simulations were correct.

REFERENCES

Gary Klein (1998), *Sources of Power, How People Make Decisions,* **MIT Press**

Until the 1990's essentially all research on decision making was done in college laboratories. The subjects of the research were often college students with no special expertise in the subject they were making decision on (such expertise might bias the results). The subjects were usually not subjected to any meaningful stress. Dr. Klein pioneered research into how people with some expertise make decisions in stressful situations. The field is called Naturalistic Decision Making. Research in this very young field has already shown that much of the historic wisdom of analytical decision making is wrong, or at least is inapplicable in the real world. *Sources of Power* is the first book from the field accessible to the general public.

Gwande, A., 2009, Checklist Manifesto, Metropolitan Books

The Checklist Manifesto is a brilliant book about the power of simple checklists. Gwande shows how they have dramatically improved health care delivery in hospitals where they are used.

James Reason (1990), *Human Error,* **Cambridge University Press**

An evaluation of the causes of human error. Focused on process industries. Especially interesting for its classification of error types into skill-based, rule-based and knowledge-based errors. His description of a cognitive model consisting of attentional and automatic subsystems ties in nicely with several aspects of this course.

Susan Fisk and Shelley Taylor (1991), Social Cognition, McGraw Hill

Summary of the field of social cognition. A very easy read and a great introduction to this subject.

Karl Weick, Kathleen Sutcliff (2007), *Managing the Unexpected, Resilient Performance in an Age of Uncertainty, Second Edition,* **John Wiley and Sons**

In the 1980's researchers at U. California at Berkley studied organizations that operated hazardous technology, yet seemed to have fewer than their fair share of accidents. In this book, Weick and Sutcliff apply insights from their study of sensemaking to make sense of these organizations. The result is a characterization of Highly Reliable Organizations (HROs).

Alistair Cockburn (2001), Writing Effective Use Cases, Addison-Wesley

A description of how to write use cases. Use cases provide a structured format for describing the process of using items. They provide an excellent format for developing operating procedures.

H.C. Sayre (1994), "The HAZOP Technique as a Structure for Operator Training", SPE 27271, Presented at Society of Petroleum Engineers Second international Conference on HS&E, Jakarta, Indonesia, Jan 25-27, 1994 (available @ spe.com)

Howard Duhon and Jorge Garduño, (2014), "Standard Operating Procedures as a Catalyst for Culture change", SPE-170824, Presented at SPE Annual Conference and Technology Exhibition, Amsterdam, The Netherlands, 27-29 October 2014

Howard Duhon, Jorge Garduño, Noel Robinson, (2009) "Planning and Procedures for Initial Startup of Subsea Production", SPE-123790, Presented at SPE Annual Conference and technology Exhibition, New Orleans, LA, USA, 4-7 October, 2009

APPENDIX 1:

Potable Water System SOP Example

Potable Water Process Description

The Potable Water System is designed to produce, store and distribute clean potable water throughout the platform. The Jockey Pumps transfer seawater to the Potable Water System where it is purified through a series of filters and membranes prior to storage and distribution to users.

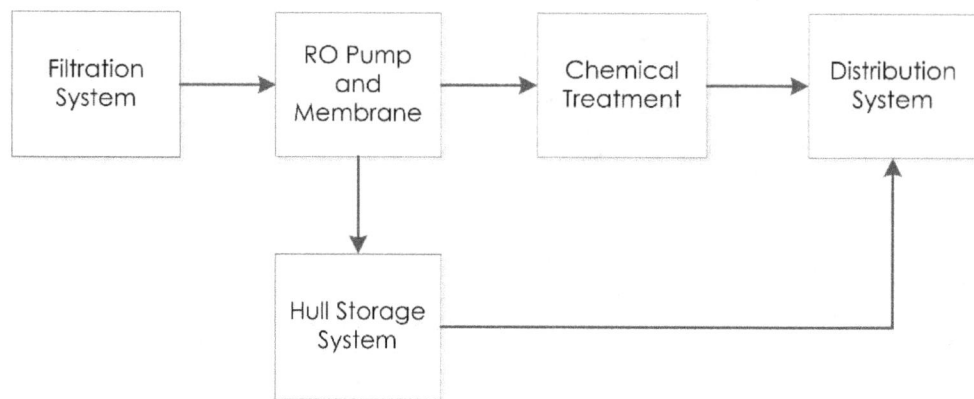

The design seawater feed flowrate to the potable water system is 11.3 GPM.

Filtration

The seawater enters three backwashable Media Filters MAJ-1010/1011/1012. The Media Filters remove the large suspended solids in the water. These are equipped with pneumatically operated backwash valves sequenced by the PLC Controller.

Backwashing occurs automatically on a set schedule (operator adjustable) but can also be started manually at the control panel. During backwashing, water is up-flowed through the Media Filters at 12 GPM to overboard. The three tank design accommodates backwashing without interrupting the filtration process.

Downstream of the Media Filters, Carbon Tank, MAJ-1015, removes chlorine. Chlorine removal is important as chlorine will foul the membranes.

The seawater then flows through 10 micron bag filters, MAJ-1016 and 1017, and 5 Micron cartridge filters, MAJ-1018 and 1019, where small suspended solids are removed. Differential pressure gauges across the cartridge filters are provided.

Reverse Osmosis – RO Pump and Membranes

Following filtration, the seawater enters the suction of the RO Pump PBE-2000 which boosts the seawater pressure upstream of the RO Membrane Pressure Vessels. The Membrane Pressure Vessels house thin-film composite spiral wound membrane elements. These separate dissolved solids (such as salts) from the seawater at an ionic level producing potable water.

A differential pressure of 800 to 1000 PSI will be maintained across the membrane elements by Back Pressure Valve FV-2025 located on the concentrate outlet from the Membrane Pressure Vessels.

Automatic shutdown switches for RO Pump PBE-2000 are incorporated for low pump suction pressure and high pump discharge pressure. The low pressure switch setting is 29 PSI (decreasing) and the high pressure switch setting is 1000 PSI (increasing). The low suction pressure or high discharge pressure indicator lights will illuminate when pertinent.

Downstream of the RO Membrane Pressure Vessels, conductivity analyzer AE-2025 measures the total dissolved solids (TDS) content and transmits signal to the PLC Controller. The PLC Controller operates 3-way valve, FV-2030, sending water with TDS content greater than 500 PPM (bad water) overboard and water with TDS content less than 500 PPM (good water) to storage.

Chemical Treatment

Two chemical treating steps, chlorine addition and mineral addition (calcite), exist downstream of the RO membranes.

Chlorine is injected to kill bacteria. A Chorine Injection Pump (PBA-2040) and 15 gallon Chlorine Tank (ABJ-2045) are located downstream of the membranes. The Chlorine Injection Pump is set to continuously inject a chlorine solution stored in the Chlorine Tank to achieve a concentration of 1 PPM in the potable water stream. Chlorine concentration is measured manually.

From there, the water flows through the Neutralization Media Tank (MBA-2050). The Neutralizing Media Tank is filled with Calcite, a mineral which is high in calcium carbonate. When the potable water flows through this media it dissolves some of the calcium carbonate increasing the pH. This serves to balance the pH of the product water and reduce corrosion in the downstream piping.

Product potable water flow is measured and displayed continuously at flow meter FQI-2050. The design potable water generation rate is 3.47 GPM or 5000 GPD.

Storage and Distribution

The potable water is then routed to storage

[The *Storage and Distribution System is not included in this example.*]

Membrane Cleaning

A Cleaning Solution Tank, MBJ-2035, is installed for the periodic in-place cleaning of the membrane elements. The membranes will become fouled by mineral scale, biological matter, colloidal particles, and insoluble organics. Deposit build-up on the membrane surface during operation will cause a loss in product water flowrates or an increase in the TDS content. As such it is recommended that membrane cleaning operation be performed at least every 30 days or when:

- Product water flow decreases 15% below the normalized reading.

- Conductivity reading increases 15% above the normalized reading.

- Membrane pressure drop increases 15% above the normalized reading.

The Cleaning Solution Tank (MBJ-2035) has a capacity of 15 gallons and will be filled with potable water from the Potable Water Distribution System. This water is routed through Carbon Filter (MAJ-2030) just upstream of the Cleaning Solution Tank. The Carbon Filter serves to remove chlorine that could foul the membranes.

Water and cleaning solution are circulated from the Cleaning Solution Tank through the Membrane Pressure Vessels to overboard by the Membrane Cleaning Pump (PBE-2036). A membrane cleaning flowrate of 15 GPM and membrane cleaning pressure of 100 PSI are achieved by the Membrane Cleaning Pump during cleaning operations.

System Sketch

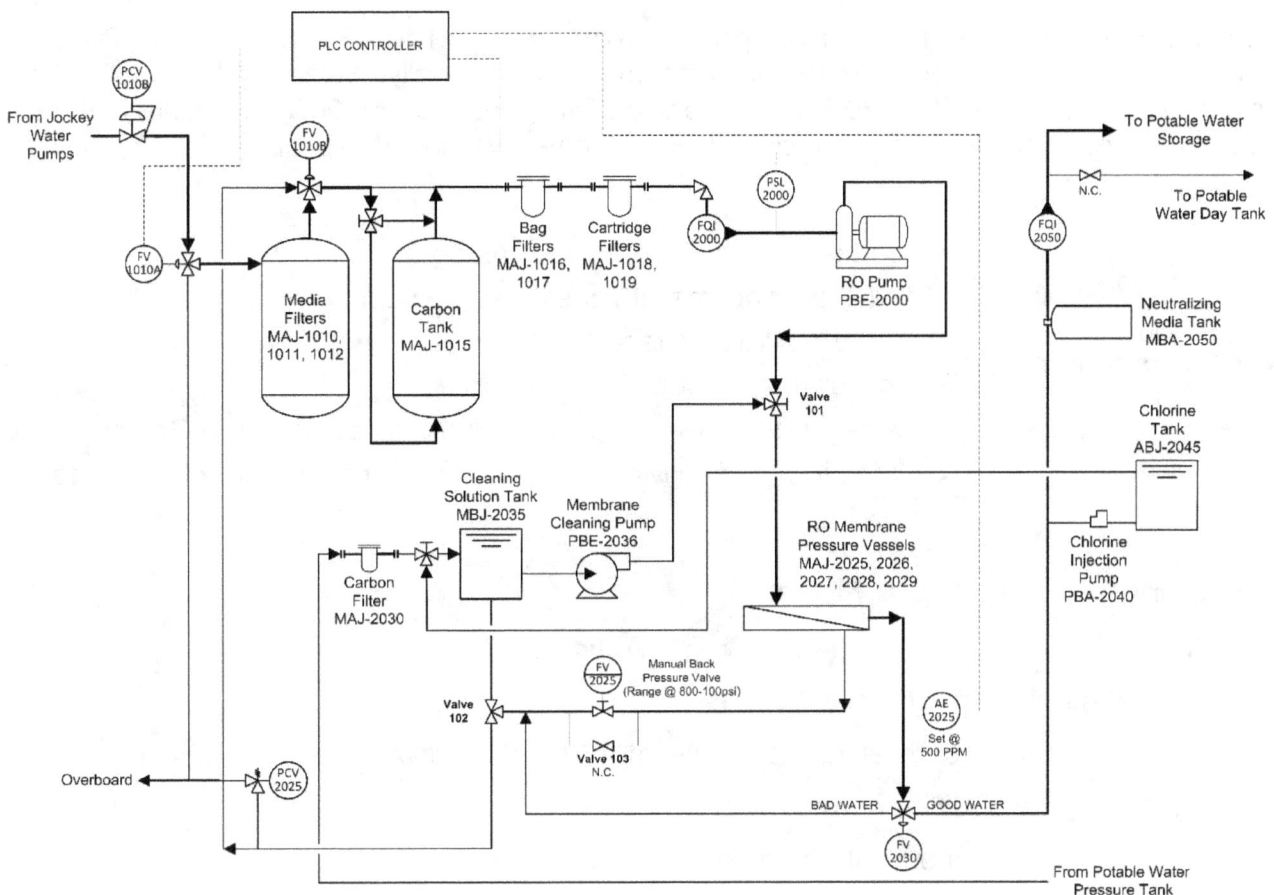

Startup Procedure - Potable Water Maker

The RO Pump will be started to pump seawater through the RO Membrane Pressure Vessel, where dissolved solids are separated out. Initially, water will be directed overboard by valve FV-2030. When conductivity analyzer AE-2025 measures a total dissolved solids content of less than 500 PPM, valve FV-2030 will divert water to potable water storage.

Assumptions	<ul><li>Jockey Water Pumps are running</li><li>All filters and membranes are installed</li><li>System is liquid filled</li><li>Instrument Air Supply is available</li><li>Adequate level of chlorine solution is available in Chlorine Tank</li><li>The Neutralizing Media Tank is adequately filled with calcite</li></ul>

Startup Potable Water Maker		
No.	**Action**	**Tick**
A. Preliminary Checks and Actions		
1.	Confirm seawater inlet pressure at PI-1010B is between 60 and 80 PSI	
2.	Align Valves for normal operations	
3.	Confirm instrument air pressure of 100 PSIG at skid	
4.	Calibrate/Check Analyzer Element AE-1025	
5.	Bleed air from RO Pump	
B. Set Control System for Startup		
6.	Turn Mode Selector Switch to the "NORMAL" position	
7.	Turn Off-On-Start Switch to the "OFF" position to reset and clear all safety shutdown conditions.	
8.	Turn Off-On-Start Switch to the "ON" _TDS meter will light. (If other indicator lights are ON, refer to Maintenance Manual.)_	
C. Start RO Pump PBE-2000		
	The Water Quality Meter will read 9999 (full scale) when the system is first started. After a couple of minutes of operation under pressure, the PPM reading should slowly begin to decrease. _When the meter reading reaches 500 PPM, the Bad Water Light will go off and valve FV-2030 will divert potable water through the product water flow meter, indicating fresh water is being delivered to the Potable Water Tanks. The PPM reading will continue to decrease until membrane rejection stabilizes._	

Startup Potable Water Maker

No.	Action	Tick
9.	Turn Off-On-Start Switch to the "START" position, hold for 2 seconds and release. *Switch will return to "ON" position. The RO Pump will start and system operation will begin.*	
10.	Confirm feed flowrate to pump is steady at FQI-2050 *If Feed Water Flow meter reading fluctuates, refer to Maintenance Manual.*	
11.	Adjust Manual Back Pressure Valve FV-2025 downstream of the Membrane Pressure Vessels until the Pump Outlet Pressure Gauge reads 800 PSI	

D. Increase the flowrate by increasing the membrane backpressure.

Increasing the Manual Back Pressure will increase the RO Pump discharge pressure as well as the flowrate through the Membrane Pressure Vessels.

Caution: Do not exceed a RO Pump discharge pressure of 1000 PSI. A discharge pressure reaching or exceeding 1000 PSI will trigger the high pressure switch and shutdown the potable water maker. This is to protect the membranes, pump, and piping from over pressure.

No.	Action	Tick
12.	Adjust the manual backpressure valve as needed until the desired product water flowrate is obtained at flowmeter FQI-2050.	

E. Start Chlorine Injection Pump PBA-2040

No.	Action	Tick
13.	Monitor Chlorine concentration. Adjust pump flow to achieve 1 ppm concentration.	

F. Monitor Operations

No.	Action	Tick
14.	Confirm incoming water pressure at PCV-1010B is set to 80 PSI	
15.	Monitor pressure drop across filters *Pressure Drop across filters should be:* <ul><li>*Media Filters < 15 PSI*</li><li>*Bag Filters < 15 PSI*</li><li>*Cartridge Filters < 15 PSI*</li><li>*Carbon Filter < 10 PSI*</li></ul>	
16.	Confirm RO Pump suction and discharge pressure	
17.	Monitor water quality via AE-2025	
18.	Confirm backwash occurs automatically as scheduled *The Backwash cycle indicator light will illuminate during backwash sequence* *Manual start of back wash cycles can be performed via push button located at the control box*	
19.	Monitor level in Potable Water Tanks	

Startup Potable Water Maker		
No.	**Action**	**Tick**
20.	Monitor the pH level of the product water downstream of the Neutralizing Media Tank *This should be checked once a month to assure proper pH levels. A decrease in pH level indicates that the Neutralizing Media Tank needs to be refilled with calcite.* *The calcite is continuously consumed and will need to be refilled after several months of continuous operation.*	
21.	Monitor membrane performance *When a sharp decline in process water flow or sharp increase in the conductivity reading is detected membranes may require cleaning* *Clean membranes at least once every month or as required.*	

Integrity Tests/Inspections

Potable Water System		
Equipment	**Integrity Test/Inspection**	**Frequency**
Quarters building	Obtain potable water samples for test *Tests to include:* • *Coliform contamination* • *Chlorine level*	Quarterly
Quarters building	Obtain potable water samples for BTEX test *Tests to include:* • *Coliform* • *Nitrate* • *Benzene* • *Toluene* • *Ethyl Benzene* • *Xylene* • *Lead*	Annual

Preventive Maintenance

Potable Water System		
Equipment	**Integrity Test/Inspection**	**Frequency**
R/O water maker	Perform and record system checks *Checks to include:* • *Equipment performance* • *Pressures* • *Conductivity test to verify TDS level* • *PH and free chlorine* • *Water levels in storage tank* • *Corrosion inhibitor tank level and pump rate* • *Chlorine tank level and pump rate*	Daily
R/O Water maker	Flush and inspect strainer	
R/O Water maker	Change media filters and charcoal beds	6 Month
Pump PM	Perform pump PM *Pumps to include:* • *R/O water maker* • *Membrane cleaning pump* • *SW & SE hull potable water pumps* • *Storage water pump*	6 Month/ Annual
R/O unit	Low pressure switch test	Quarterly